CAMBRIDGE LIBRARY COLLECTION

Books of enduring scholarly value

Life Sciences

Until the nineteenth century, the various subjects now known as the life sciences were regarded either as arcane studies which had little impact on ordinary daily life, or as a genteel hobby for the leisured classes. The increasing academic rigour and systematisation brought to the study of botany, zoology and other disciplines, and their adoption in university curricula, are reflected in the books reissued in this series.

Practical Hints Upon Landscape Gardening

William Sawrey Gilpin (1761/2–1843), landscape painter and illustrator, later became a landscape gardener and writer. He set himself up as a drawing master in Paddington Green and also illustrated picturesque travel-writing. Between 1804 and 1806 he was the first president of the Society of Painters in Water Colours, and then the third drawing master at the Royal Military College in Marlow. After being discharged from this post, Gilpin became a successful landscape gardener and advisor to the nobility. His approach to landscape gardening was influenced by painting and Sir Uvedale Price's *Essay on the Picturesque* (1794). Gilpin's *Hints*, published in 1832, advocates that landscapes should be improved by the 'taste' of a painter's eye, and artificial buildings united with their surroundings. Like his landscape practice, this book was highly regarded by Gilpin's contemporaries for its emphasis on the picturesque, especially when landscape gardening centred upon the introduction of exotic plants.

Cambridge University Press has long been a pioneer in the reissuing of out-of-print titles from its own backlist, producing digital reprints of books that are still sought after by scholars and students but could not be reprinted economically using traditional technology. The Cambridge Library Collection extends this activity to a wider range of books which are still of importance to researchers and professionals, either for the source material they contain, or as landmarks in the history of their academic discipline.

Drawing from the world-renowned collections in the Cambridge University Library and other partner libraries, and guided by the advice of experts in each subject area, Cambridge University Press is using state-of-the-art scanning machines in its own Printing House to capture the content of each book selected for inclusion. The files are processed to give a consistently clear, crisp image, and the books finished to the high quality standard for which the Press is recognised around the world. The latest print-on-demand technology ensures that the books will remain available indefinitely, and that orders for single or multiple copies can quickly be supplied.

The Cambridge Library Collection brings back to life books of enduring scholarly value (including out-of-copyright works originally issued by other publishers) across a wide range of disciplines in the humanities and social sciences and in science and technology.

Practical Hints Upon Landscape Gardening

With Some Remarks on Domestic Architecture,
as Connected with Scenery

WILLIAM S. GILPIN

CAMBRIDGE
UNIVERSITY PRESS

CAMBRIDGE UNIVERSITY PRESS

Cambridge, New York, Melbourne, Madrid, Cape Town,
Singapore, São Paolo, Delhi, Mexico City

Published in the United States of America by Cambridge University Press, New York

www.cambridge.org
Information on this title: www.cambridge.org/9781108055642

This edition first published 1832
This digitally printed version 2013

ISBN 978-1-108-05564-2 Paperback

Selected books of related interest, also reissued in the
CAMBRIDGE LIBRARY COLLECTION

Amherst, Alicia: *A History of Gardening in England* (1895) [ISBN 9781108062084]

Anonymous: *The Book of Garden Management* (1871) [ISBN 9781108049399]

Blaikie, Thomas: *Diary of a Scotch Gardener at the French Court at the End of the Eighteenth Century* (1931) [ISBN 9781108055611]

Candolle, Alphonse de: *The Origin of Cultivated Plants* (1886) [ISBN 9781108038904]

Drewitt, Frederic Dawtrey: *The Romance of the Apothecaries' Garden at Chelsea* (1928) [ISBN 9781108015875]

Evelyn, John: *Sylva, Or, a Discourse of Forest Trees* (2 vols., fourth edition, 1908) [ISBN 9781108055284]

Farrer, Reginald John: *In a Yorkshire Garden* (1909) [ISBN 9781108037228]

Field, Henry: *Memoirs of the Botanic Garden at Chelsea* (1878) [ISBN 9781108037488]

Forsyth, William: *A Treatise on the Culture and Management of Fruit-Trees* (1802) [ISBN 9781108037471]

Haggard, H. Rider: *A Gardener's Year* (1905) [ISBN 9781108044455]

Hibberd, Shirley: *Rustic Adornments for Homes of Taste* (1856) [ISBN 9781108037174]

Hibberd, Shirley: *The Amateur's Flower Garden* (1871) [ISBN 9781108055345]

Hibberd, Shirley: *The Fern Garden* (1869) [ISBN 9781108037181]

Hibberd, Shirley: *The Rose Book* (1864) [ISBN 9781108045384]

Hogg, Robert: *The British Pomology* (1851) [ISBN 9781108039444]

Hogg, Robert: *The Fruit Manual* (1860) [ISBN 9781108039451]

Hooker, Joseph Dalton: *Kew Gardens* (1858) [ISBN 9781108065450]

Jackson, Benjamin Daydon: *Catalogue of Plants Cultivated in the Garden of John Gerard, in the Years 1596–1599* (1876) [ISBN 9781108037150]

Jekyll, Gertrude: *Home and Garden* (1900) [ISBN 9781108037204]

Jekyll, Gertrude: *Wood and Garden* (1899) [ISBN 9781108037198]

Johnson, George William: *A History of English Gardening, Chronological, Biographical, Literary, and Critical* (1829) [ISBN 9781108037136]

Lindley, John: *The Theory of Horticulture* (1840) [ISBN 9781108037242]

Knight, Thomas Andrew: *A Selection from the Physiological and Horticultural Papers Published in the Transactions of the Royal and Horticultural Societies* (1841) [ISBN 9781108037297]

Loudon, Jane: *Instructions in Gardening for Ladies* (1840) [ISBN 9781108055659]

Mollison, John: *The New Practical Window Gardener* (1877) [ISBN 9781108061704]

Paris, John Ayrton: *A Biographical Sketch of the Late William George Maton M.D.* (1838) [ISBN 9781108038157]

Paxton, Joseph, and Lindley, John: *Paxton's Flower Garden* (3 vols., 1850–3) [ISBN 9781108037280]

Repton, Humphry and Loudon, John Claudius: *The Landscape Gardening and Landscape Architecture of the Late Humphry Repton, Esq.* (1840) [ISBN 9781108066174]

Robinson, William: *The English Flower Garden* (1883) [ISBN 9781108037129]

Robinson, William: *The Subtropical Garden* (1871) [ISBN 9781108037112]

Robinson, William: *The Wild Garden* (1870) [ISBN 9781108037105]

Sedding, John D.: *Garden-Craft Old and New* (1891) [ISBN 9781108037143]

Veitch, James Herbert: *Hortus Veitchii* (1906) [ISBN 9781108037365]

Ward, Nathaniel: *On the Growth of Plants in Closely Glazed Cases* (1842) [ISBN 9781108061131]

For a complete list of titles in the Cambridge Library Collection please visit:
http://www.cambridge.org/features/CambridgeLibraryCollection/books.htm

PRACTICAL HINTS

UPON

LANDSCAPE GARDENING:

WITH SOME REMARKS

ON DOMESTIC ARCHITECTURE,

AS CONNECTED WITH SCENERY.

BY

WILLIAM S. GILPIN, Esq.

LONDON:

PRINTED FOR T. CADELL, STRAND;

AND W. BLACKWOOD, EDINBURGH.

1832.

LONDON:
Printed by A. & R. Spottiswoode,
New-Street-Square.

CONTENTS.

A 2

CHAP. VI.

MISCELLANEOUS.

INTRODUCTION.

This little work may probably, at first sight, appear superfluous to those who have read the Essays of Sir Uvedale Price upon this department of taste; where the subject is so ably and so fully discussed as to leave no room for improvement, no ground for dissent. Still, however, notwithstanding the extended spirit of improvement in Landscape Gardening, it may be presumed that numbers have not read those Essays; whilst others, from the want of previous knowledge on the subject, may not be able to reap the full information they contain, so as themselves to direct, upon the principles of taste, the improvements they desire, or to appreciate the ability of others to whom they intrust a work of no light interest, either to the owner of the place, or as connected with the general diffusion of taste through the country at large.

Indeed, Sir Uvedale Price's preface to the

first volume of his second edition justifies
the utility of such a work. I beg to make
the following short extract : ——

" As the general plan and intention of my
" first publication," says Sir Uvedale, " have
" been a good deal misunderstood, I wish to
" give a short account of them both. The
" title itself might have shown, that I aimed
" at something more than a mere book of
" Gardening. Some, however, have conceived
" that I ought to have begun by setting forth
" all my ideas of lawns, shrubberies, gravel
" walks, &c.; and, as my arrangements did
" not coincide with their notions of what it
" ought to'have been, they seem to have con-
" cluded, that I had no plan at all."

What Sir Uvedale here leaves his readers
to gather from the whole of his interesting
and instructive work, it is the aim and intention of the following pages (as far as
relates to the immediate subject of Landscape
Gardening) to concentrate, and to render
practically useful.

All the writers on this subject that I have
met with (the author of the Essays except-
ed), whatever be their comparative merit,
appear to me to be more or less defective

in practical information. The author of
" Design in Gardening *" accuses the " Ob-
" servations on Modern Gardening " of this
omission; and though I have carefully read
both the works, I must confess myself to have
found as little of practical information in the
former as in the latter author, and far less of
interest and taste.

It will be remembered, that the authors I
mention were none of them professional improvers. Their observations, therefore, however interesting they may be to those who
are conversant with the subject, will be deficient in that general utility and practical
information which is the object of the following pages; the merit of which, if they have
any, will consist in opening the general principles of taste to those who have not studied
the subject, and in thus enabling them to appreciate each the character of his own place,
and the different schemes that may be suggested for its improvement, will afford a
source of increasing variety and delight.

Agreeing fully with Sir Uvedale Price in his
estimate of the requisites necessary to form
a just taste in Landscape Gardening, I am

* Essay on Design in Gardening, p. 149.

emboldened to submit to the public my ideas upon the subject, having been bred to the study of Landscape Painting in the first instance, and having for many years applied the principles of painting to the improvement of real scenery.

It has ever appeared to me, that a very essential part of an improver's duty is to explain to the proprietor the principles upon which he suggests any plan of improvement. This, during the progress of the work, not only enhances the pleasure of the proprietor, and assists his general taste, but it also ensures his future care, through the periodical prunings and thinnings which must of necessity take place, that the original scheme of the improvement be kept in view. It will also frequently happen that local circumstances, or individual prejudices may be opposed to the plan of improvement recommended. In such cases, I have generally found, that a full explanation of the principles on which the plan is founded will not fail to overcome those prejudices, and modify such local circumstances, so that they shall not materially interfere with the general design. If the improver understands his profession, such a discussion must be highly desirable to him, whatever be the result.

Taste, as connected with general feeling, is more or less subject to the influence of fashion. We perceive this influence in dress, ornament, plate, &c., as well as in architecture and gardening; and, as alteration usually ends in extremes, so within the last century taste has experienced the sweeping hand of reform. Simplicity became the standard of the day: and, as the richly embossed plate of former times was superseded by the bald and meagre productions of more modern simplicity; so the ample terrace, with its massive balustrade, its steps, fountains, and alcoves, with all its rich, though formal, accompaniments of parterres, backed by the sheltering skreen of venerable evergreens, fell beneath the indiscriminating hand of reform, and left the mansion stripped of those embellishments which time had, as it were, identified with its very existence, to lament over the insipid simplicity and baldness spread around it.

Time and reflection seem at length to have enabled us to judge with impartiality between the old and new systems; and the principles of taste are, from various causes, better understood, and more generally diffused, than at any former period.

In the article of plate, for instance, the richness of the old is imitated in the modern manufacture, whilst the former is itself sought after with avidity. So on the subject of this discussion, the same improvement seems to be taking place; and richness, intricacy, and variety, have entered the lists against insipidity, distinctness, and dull uniformity. The bold, though formal, stretch of terrace ventures occasionally to re-occupy the situation from which the easy curve had almost universally ejected it; and we may hope the time is approaching, when Sir Uvedale Price's prophecy will be accomplished in the union of the excellencies of the two systems.

As the embellishments that surround the country residences of England are extended over a much wider range than formerly, their influence on the general character of the country must be proportionably increased. It is highly desirable, then, that these embellishments should be founded on the principles of true taste; which, as the Essays before alluded to have abundantly proved, is only to be perfected by the united study of nature and the works of the best landscape painters. A taste, thus formed, can alone produce that variety which the natural character of each place will suggest to an eye conversant with the principles of composition; whilst he, who is unacquainted with those principles, must be in danger of repeating the same scheme of operation, with little or no relation to the character of the different places to which it is applied.

The object of the following pages being (as I have already stated) to suggest a few leading hints, whereby, at least, the great outline of taste may be preserved, it will be necessary to accommodate these hints to places of various sizes; for the hand of taste may be discovered in the embellishment of half an acre, though the want of it will not be so offensive as on the more extended scale of a pleasure ground or a park. This diversity of application will unavoidably create occasional repetition of such remarks as are equally suited to places of different extent.

In order to render the principles here suggested more practically useful, a few illustrative plans and sketches are added, in which utility alone has been attended to, as any thing beyond that would be an useless addition of expense.

Painsfield, East Sheen,
April 7. 1832.

ERRATA.

Page 62. line 14. for " prefesable," read " preferable."
85. line 12. for " were," read " where."
108. line 14. for " predilecton," read " predilection."
114. lines 3, and 4. for " Empiricsin," read " Empirics in."

PRACTICAL HINTS

ON

LANDSCAPE GARDENING,

&c.

CHAPTER I.

GENERAL IMPROVEMENT. — FORMATION. — REMOVAL. — SITUATION FOR A HOUSE. — CHARACTER OF THE HOUSE. — DIVISION OF SCENERY.

THE judicious improvement of any place must rest upon the natural or acquired character of the place itself. I say *acquired* character, because many places may be found where the natural character has been superseded by planting, and other decorations of such long standing as forbids their removal, and directs future improvement to harmonise with the existing state of things.

Improvement may be classed under two

B

leading heads, *formation* and *removal:* the
former will be more especially requisite in the
decoration of a new place; the latter, in the
correction of an old one. I will consider, first,
the formation of a new place.

When a house is to be built, its situation,
size, and character are well worthy of consideration, as connected with the general harmony
of the scene.

Though it would be difficult to fix the pre-
cise situation of a house without seeing the
ground over which it is to preside, yet a few
hints may be given that will, at least, prevent
the glaring errors we too often witness on this
head, as also on the character of the house
itself, and more particularly on the approach
to and entrance into it.

I will consider the situation of a house in
relation to the extent of the domain; the
immediate ground on which it is placed;
the scenery it commands; the shelter requisite to comfort within it; and the access
to it.

A house ought rarely to be placed on the
highest point of the domain, if that point is
of any considerable height; as I conceive it
would be objectionable to most of the requi-

sites just mentioned: for, though circumstances will sometimes demand an elevated
situation, yet it is by no means favourable to
comfort or access; nor is the general composition usually so good as from a lower station.

I consider a house to be best situated when
it stands upon a platform, with a rising ground
behind it, and with a depth below it: no position can be more favourable for that variety
of embellishment so desirable around the
house. When the undulation of ground is
of a more gentle character, I would still fix
my house as nearly upon this plan as circumstances might admit. With regard to the
scenery from the house, I should be careful to
get, if possible, from my windows, some large
trees for a foreground, as essential to the
general composition; a point of much more
consequence than a mere extensive view.

I think it a great mistake in the placing of
a house, to set it parallel to a river or a valley.
I remember a house so placed with regard to
the Thames, in a very beautiful part of its
course; but, from the situation of the house,
you look straight across the river, which nar-
rows it to little better than a canal, whilst the
reaches up and down the stream, though both

are beautiful, are altogether lost to the windows.

There is an universal and well founded preference of a south aspect for the living rooms of a house: but if, by yielding a few points of the compass on either side, I could improve the composition, I confess I should not hesitate so to do. When the house is of an irregular form, windows may be so placed as to command a greater variety of scenery than can be obtained by the usual rectangular building. I remember seeing a house built by the late Colonel Mitford, near the New Forest, in a triangular form, to meet three distinct and very different views. The situation of a house will, in a great degree, be determined by its character; which character will, again, be mainly influenced by that of the surrounding scenery, particularly of its immediate domain.

Country residences, such as we are treating of, may be classed under the following denominations: a Castle; a Grecian or Italian Edifice; a Manorial Building; a Hunting Lodge; and a Cottage ornée.* The distinction

* A general name for that sort of building that claims no place among the former.

between the two last will, perhaps, depend as much upon the scenery in which they are placed, as upon any essential variety in the buildings themselves.

Though situations may sometimes occur where the choice of an appropriate building may not be so obvious, yet I conceive that, in many others, good taste could be at no loss in making a judicious selection; and, even in cases not so clear, the same good taste would avoid any glaring disunion between the house and its accompanying scenery.

In speaking of character in scenery as connected with our present purpose, I will venture to range it under five heads: for though, in many instances, the Romantic and the Picturesque may appear to blend into one character, yet, in very many others, a marked distinction will, I conceive, be found between them. The Romantic will, perhaps, often include the Picturesque; but, in numerous instances, I think, the latter will be found unaccompanied by the former quality. I will also add, as connected with residences, the rural, as a distinct class from any of the rest. Scenery, therefore, may be divided into the

Grand, the *Romantic*, the *Beautiful*, the *Picturesque*, and the *Rural*. The first is characterised by largeness and unity of parts: its contrasts are few and bold. Such is the scenery, generally, of a mountainous country, and more especially of the sea, when viewed from a commanding station. Lake scenery usually comes under this character. The view from the ridge of the Cotswold Hills over the vale of Severn well deserves this title; as does also, though of a different composition, the view from the house at Brockenhurst, in the New Forest, where the eye sweeps over a mass of majestic wood, apparently interminable, until it melts into the horizon.

Romantic scenery is wrought upon a smaller scale than the former, with more parts, and a greater variety and quickness of transition from part to part. It is marked by precipitous steeps; angular rocky projections forcing their way between the rugged stems that are rooted in their crevices, or rising out of the wild undergrowth at their base. Intricacy seems the leading feature of the Romantic. The Grand bursts at once upon the eye, and holds it in astonishment: the romantic leads you onward in alternate expectation and discovery,

whilst Tranquillity appears the presiding genius of the scene.

Such, I should say, is the character of
Dovedale, in Derbyshire, and, upon a smaller
scale, of Corby and of Nunnery, in Cumber-
land, as also of Rokeby. The scenery of
Bolton Abbey on the Wharf is a fine specimen of the romantic.

The Beautiful in scenery is characterised
by more gentle contrasts, with broader folds
of ground, and smoother surface; whilst its
embellishment consists in groups of trees of
ample growth and erect stature. Where water
is added, you have all the requisites of the
Beautiful. Longleat, Bowood, and Marston,
amongst many others, are good examples of
this character.

The Picturesque scene is marked by smaller
and more abrupt folds of ground, with but
little of flat surface, and clothed in a rougher
mantle. Its wood is usually of less ample
growth, and mixed with thorns, hollies, gorse,
broom, brambles, &c. This description of
country is frequent in some parts of Kent;
and, perhaps, Seven Oaks Common may be
selected as an example very generally known.
Holwood, in that neighbourhood, comes under

this character: Addington Park, near Croydon, is also a striking specimen of the Picturesque.

The Rural comprehends a large portion of the scenery of many counties in England. Several very pleasing examples are to be found in Surrey, which, as far as I have had opportunities of judging, contains the highest class of this description of country. Undistinguished by any great features, its power of pleasing depends principally upon its hedge-row timber, producing the appearance of a well wooded county, softening gradually into a rich distance. Open commons here and there give an interesting variety to the general mass of cultivation; though, in this point of view, it is to be lamented that the increased spirit of agriculture has much curtailed this prominent feature of rural landscape.

It will be evident to every observer of nature, that these different characters are subject to various occasional combinations; which, nevertheless, though they may lessen, do not destroy the distinction. The road between Epsom and Dorking furnishes a striking example of the union of the Rural with the Beautiful in scenery: and it would be difficult to name a drive of such pleasing interest, where the bold-swelling hills, crowned with their decorated mansions, gradually descending into the wooded vale, enlivened by village and hamlet of peculiar neatness, form altogether a combination of beauty, richness, and comfort, which at the same time delights the eye, and awakens the mind to a train of interesting reflections.

This occasional mixture of character in the scenery will naturally influence the character of the building to be erected upon it; other circumstances, also, will have their due weight on the question: moreover, it will be remembered that hints, and not rules, are here suggested, with a view of preventing the more flagrant violations of harmony between the house and the scenery round it.

In adapting, then, a mansion to a grand situation, the choice of building would, I conceive, be principally influenced by the character of the immediate ground on which the house was to be placed. If that consisted of gentle undulations, with sufficient extent of lawns, shrubberies, &c., I should prefer a Grecian elevation: if, on the contrary, the site for the building were of limited extent and abrupt character, I should esteem it bet-

ter suited to the Castle.   Ardgowan, in Ayr-
shire, and Kinfauns, near Perth, will illustrate
this observation.

The former, though of considerable height,
yet, being approached by a gradual ascent
through easy swelling folds of ground, might
have been properly crowned with a Grecian
Edifice, had the immediate ground on which
it stands been of sufficient extent and easy
surface.   As it is, I think the character of
the mansion the only fault, in a place where
grandeur and variety are more happily
blended than I have any where met with.

Through an opening in the wood, which
clothes the south side of the eminence, you
catch a little bay of the Clyde, enlivened by
all the circumstances of fishing boats, figures,
nets, &c., combined with the straggling
skirts of the village, and backed by a bold
swell of hill.   Looking more to the west, you
have the Isle of Bute, with the romantic
peaks of Arran rising behind it, and the sea
extending beyond them to the Irish coast;
whilst following the prospect round to the
north, the Clyde, from the more contracted
line of the opposite shore, assumes the character of a magnificent lake, stretching its vary-

ing reaches up to Loch Long, and bounded by the grand mountain line of Loch Lomond.

The situation of Kinfauns, being of greater height and very abrupt, is properly occupied by a Castle, with its bold terrace overhanging the steep, up which the approach ascends to the entrance, having wound round the edge of the height under the shade of a venerable row of trees. Having gained the Castle, the eye breaks at once upon a splendid view of the Tay winding its broad course below, occasionally interrupted by the tops of majestic wood hanging on the steep, and enriched with a variety of vessels passing and repassing to and from Perth.

Had the abrupt knoll on which the castle stands been planted with thorn, holly, juniper, &c., so as to form a mass of undergrowth below the terrace wall, the effect of the whole would have been perfect: the want of this gives a newness and poverty of character ill suited to the general richness of the scene.*

To Romantic scenery also the Castle seems well adapted. Its angular projecting buttresses, its towers of irregular height, its walls

* Having strongly urged this opinion to the noble owner, I trust there is no impropriety in stating it here.

incorporated with the rocky steep on which it stands, are all in harmony with the scene. But, should the situation afford the choice of a less abrupt site with equal advantages, I should prefer the Manorial House: for, though a Castle is no longer connected with alarm, yet the Manorial Residence is more strictly in unison with that soothing tranquillity which pervades the Romantic scene.

Bardon Tower, which is still standing on a height commanding the windings of the romantic valley of Bolton Abbey, appears as if it had been the former residence of the place; though, in fact, it was not so. Were a house now to be erected, I should wish to place it where the river, having forced its agitated way through a rocky channel of three miles, spreads itself into a little tranquil lake, gently winding round its varying shore, till its stream, gradually entering a more confined channel, glides silently through the woody scenery below.

On the border of this little lake would I place my house,—where, indeed, the good taste of former times has placed the Abbey and where a corresponding taste has fixed the Vicarage, which evinces the eye that placed it to be worthy of the enchanting scenery with which it is surrounded.

The Grecian Edifice is best suited to preside over the scenery designated by the term *Beautiful.* Its regular proportions and high-finished decorations are in unison with the soft and polished character of all around it, where elegance and gaiety hold unlimited sway.

Here also may, with propriety, be placed the Manorial Building; only I would (if circumstances permitted) set it deeper in its woody back ground than I would the gayer Grecian Mansion, and in its embellishments aim at substituting cheerfulness for gaiety.*

The Picturesque situation seems formed for what has been termed *a Forest Lodge;* which I should describe as a building calculated rather for convenience than display: low in comparison of those before mentioned, irregular in its form, and, if the ground be favourable, in its height also; no columns, no porticos,— a porch only allowable. The pleasure

---

* Somerhill, near Tonbridge, is an exception to the general situation of the manorial house; but the splendid scenery it commands justifies the elevated station it occupies.

ground less extensive than at the last-mentioned residence, and less ornamented in its decorations.

The last description of scenery, the Rural, is calculated for the *Cottage ornée*, which, being without pretension, may be assimilated by the variety of its accompaniments to the ground it occupies, and to the scenery it commands.

Whatever be the size or character of the house under any of the above divisions, the putting the offices under ground seems to me to be a great mistake, either as it regards the appearance of the building itself, or of the ground around it. The offices may be so managed as to relieve the square box-like appearance of the house, and create a variety of height and projection in the general mass of building, which, when broken and enriched by well disposed planting, will form a much more agreeable whole, than can be produced by any single compact mass of whatever style. This is one cause of the picturesque effect of the Manorial, in which the offices are usually so placed as to give extent and variety to the pile.

On the other hand, houses, particularly of

a moderate size, frequently suffer much from the mode of attaching the offices to them ; as when they are brought on a line with the front of the house, — or, as I have seen, even projected before it,—in such case the windows of the offices on the one front command the pleasure ground, and those on the other over-look the approach: on both they materially injure the effect of the main building by excluding the return angles, and bringing the whole mass into one extended flat line. The expedient of shrubbing out the offices, as it is termed, is no improvement; as that will not restore the return angles of the main build-ing, at the same time that it forces the walk into the sight of the windows, from which it should be concealed.

The propriety of Sir Uvedale Price's re-marks upon this subject will amply apologise for my transcribing them in this place : —

" Much of the naked solitary appearance of " houses is owing to the practice of totally " concealing, nay, of sometimes burying all " the offices under ground, and that by way " of giving consequence to the mansion ; but, " though exceptions may arise from parti-" cular situations and circumstances, yet, in

" general, nothing contributes so much to
" give both variety and consequence to the
" principal building as the accompaniment,
" and, as it were, the attendance, of the in-
" ferior parts in their different gradations. It
" is thus that Virgil raises the idea of the
" chief bard, —

   " ' Musæum ante omnes ; medium nam plurima turba
      Hunc habet, atque humeris extantem suspicit altis.'

" Of this kind is the grandeur that charac-
" terises many of the ancient castles, which
" proudly overlook the different outworks,
" the lower towers, the gateways, and all the
" appendages of the main building ; and this
" principle, so productive of grand and pictu-
" resque effects, has been applied with great
" success by Vanbrugh to highly ornamented
" buildings, and to Grecian architecture. The
" same principle (with those variations and
" exceptions that will naturally suggest them-
" selves to artists) may be applied to all
" houses. By studying the general masses,
' the groups, the accompaniments, and the
" points they will be seen from, those ex-
" terior offices, which so frequently are buried,
" if not under ground, at least behind a close

" plantation of Scotch firs, may all become
" useful in the composition; not only the
" stables, which often, indeed, rival the man-
" sion, and divide the attention, but the
" meanest office may be made to contribute
" to the character of the whole, and to raise,
" not degrade, the principal part. The dif-
" ference of expense between good and bad
" forms is comparatively trifling — the differ-
" ence in their appearance immense."

When the offices required are of moderate
extent, they may be connected with the house
by a handsome screen wall, of such height as
to hide them altogether : the wall may be
partially broken by planting.

Whilst speaking of the house, I cannot
omit a circumstance, the inattention to which
has spoiled two thirds of those which I have
seen : I allude to the entrance.

In a Grecian or Italian edifice it may be
essential that the entrance should occupy the
centre of one of the fronts ; in which case I
think it equally essential that the living rooms
should not be on the same front : on the
contrary, we frequently see the entrance on
the south front, and the drawing room or
library exposed to the gaze of the servants

c

from the carriage, whilst the windows, which should have opened upon the embellishments of a terrace or a pleasure ground, look upon a sweep of glaring gravel; indeed, it is not unusual to meet with the conservatory on a line with the hall door.

I trust I shall not be deemed too severe upon this great mistake, when I state, that I have visited a house of much beauty and expense, and commanding scenery of very considerable variety and grandeur, where the library window (the only room on the south front) opens upon the approach, and the carriages drive up immediately under it: an unfortunate error, now irremediable.

Where circumstances will admit, the alteration of the entrance so misplaced is well worthy of attention. At Footscray Place the approach formerly went round the house to set down on the south side, with a flight of steps up to the hall door; the house is now entered upon the north, on a level with the hall, and the former entrance is converted into a library, having access by the flight of steps to a handsome terrace below.

As far as concerns the entrance, the irregular Manorial is preferable to the Grecian architecture; for a mansion that does not require the dimensions of a palace calls for no sacrifice in its access. Indeed, its irregularity is highly beneficial, both in the variety of the outline, and in the light and shadow resulting from it; the want of which is so obvious in the square flat surface of so many of our modern houses.

CHAP. II.

THE APPROACH.

NEXT to the situation and character of the
house, the approach to it is to be considered.

I have frequently thought that an undue
stress is laid upon the approach, as connected
with the general scenery of the place. We
often meet with it studiously carried through
some of the finest points of view, and thus
forestalling what ought to have been reserved
for the windows or the pleasure ground. The
approach to the Priory at Stanmore is an
illustration of my feeling upon this subject.
There is no doubt that the beauty of that approach, simply considered, would be improved
by the removal of a screen of high trees,
which excludes the distant country. But the
screen, notwithstanding any suggestions to
the contrary, is, with great judgment, retained, as a premature disclosure would most
materially injure the effect of the magnificent
display of scenery that bursts upon you from
the drawing-room windows.

An approach should appear to be an un-studied road to the house: —

" So let th' approach and entrance to the place
" Display no glitter, and affect no grace: " *

and its character should vary with that of the residence to which it leads. This variety will be principally marked by its length, and by its embellishments. The former of these distinctions need not always exist; the latter, I confess, I have ever held to be essential.

After breaking off from the public road, the approach should avoid any direction parallel with it, as good sense dictates the use of what is already provided, as long as it is suitable to your purpose. The inattention to this rule in places of limited extent be-trays that limitation which might otherwise escape detection. I have seen an approach running parallel with the high road, with little more than the hedge dividing them, up to the very door, and a shrubbery walk fol-lowing the same line, with scarce a wider separation between it and the approach. The lodge is a high finished temple, built, as I was

* Knight's Landscape.

c 3

informed, at the cost of three thousand pounds. I was not within the domain.

In forming the line of approach (if of any considerable length) I would avoid an uniform curve, or easy sweep, as it is termed ; there is to me a painful insipidity in a long continued curve, be it either road or walk. Where, therefore, the length is sufficient to justify deviation from the curve, I should avail myself of any fair obstacle to vary the direction of my road, which would return again at a fit opportunity to its original destination. This I take to be the idea of the poet,

" But still in careless easy curves proceed ;"

which is quite contrary to the lengthened uniform curve I have ventured to condemn. An approach not being subject to the same necessity of speed as the high road, I should seek rather than avoid any occasional undulation of ground, as conducing to the variety and interest of the scene. This, however, requires great judgment ; for a visibly needless ascent is a palpable error.

As contrast is so conducive to enjoyment, I would, by all fair means, avail myself of its aid in conducting an approach.   If the mansion commanded an extensive prospect I would take the approach through the more confined scenery, should circumstances permit it. If, on the contrary, the house reposed in a more secluded scene, I would embrace every lawful opportunity of catching from the approach those features of variety, and extent which were excluded from the house, and its immediate environs. In fine, I should endeavour (subject to what has been before advanced) to show from the approach such scenery as did not come within view from the house and the dress ground.

I have said, that the approach should vary according to the character of the residence; and that this variety will consist principally in its length and its embellishments.

There are many instances amongst the old mansions where their proximity to the high road admits of little or no approach, as at Blickling and Wilton. On the other hand, the approach of more modern times is often carried through uninteresting scenery, merely to prolong its length: where this is visible the effect is bad. Where the approach is of necessity to be carried through a length of uninteresting space, as at Clumber from Tuxford, (a distance of three miles from the outer lodge to the park gate), passing between farms in various occupation, the best way of getting over such country is by an avenue, as it is there done; which not only avoids a multiplicity of gates, but is in character with the magnitude of the domain through which it leads.

We have seen at Wilton that *length* is not always necessary, even in an approach to a magnificent residence: with regard to the other point of difference, *embellishment*, I hold it to be an essential distinction, according to the magnitude and character of the place.

By the embellishments of an approach I mean the trees and undergrowth that adorn it. These embellishments, then, ought, I conceive, to be in unison with the scene. In driving through a park interspersed with masses of wood, natural groups of trees, and thickets of thorn, holly, &c., we do not expect to meet with laurels, portugals, and other materials of a shrubbery: in all such cases I cannot but feel them utterly misplaced. The gardener has no business in the park. But, at the cottage ornée, its limited domain and general character not admitting the masses and groups of park scenery, the aid of shrubs may be allowed, restricting them, however, to the more sober classes, principally evergreens, leaving the gayer varieties to heighten the beauty and interest of the pleasure ground, properly so called. I would have no flowers, nor any thing that apparently required the gardener's care beyond neatness of keeping; let the evergreens trail upon the lawn, and no mould be seen. To the introduction of exotics in an approach of enlarged scale, I confess myself most hostile; having witnessed the approach even to a palace-like mansion carried through miles of shrubbery; and in other places have seen what is scarcely less objectionable, the approach through the wild scenery of a natural wood, spotted and disfigured by patches of shrubs and flowers. I certainly should never so decorate an approach. If I find one so treated, where time has in some degree softened the incongruity by giving freedom and ruggedness to the materials, I deal with it the best I may, judging it, in this, as in most other cases, safer to make the best of what I find, than risk the alternative of a radical reform. Sometimes, indeed, the natural character of the place will warrant the extermination of exotics so misplaced: in other situations, such a removal would materially injure the scenery, as in one of the lines of approach at Oatlands, which passes through a narrow hollow way, and where time and accident have so united the shrubs with the higher trees, that any attempt to remove them would totally destroy the beauty of the whole. Hollies, of course, are not included in the foregoing remarks, as they are the growth of the forest, as well as the ornament of the shrubbery.

It is necessary here to mention, that I consider a villa to be under the same circumstances, with regard to the approach, as the cottage ornée. Though the residence may be a palace as to size and character, yet the limited domain on which it stands is a legitimate apology for the style of its accompaniments. A villa, I conceive, can only be so termed, when within a few miles of a city; where a spacious residence is requisite, though the domain is, from circumstances, confined; but should the domain be more extensive, as at Sion House, the approach should then assume a higher character, as it there does.

The avenue, as an approach, is, in general, so destitute of composition, by cutting the l n dscape in half, that the introduction of it must depend upon the circumstances of the place itself. On the other hand, where time has invested it with dignity, and the rest of the scenery is coeval with it, temerity rather than judgment would dictate its destruction. Breaking it by partial removal is, I think, equally injudicious.

The avenue in the Phœnix Park, Dublin, is of a peculiar formation, being composed of groups of trees at regular distance from each other, and in exact line; but the groups on one side facing the openings on the other. The effect is injurious to the grandeur and solemnity of the avenue; but it gives, perhaps, a cheerfulness and variety to it as a drive, for it leads to no mansion.

CHAP. III.

DRESS GROUND. — SCENERY BEYOND IT. — COMPOSITION
IN LANDSCAPE. — FOREGROUND, NATURAL AND ARTI-
FICIAL. — OLD AND NEW SYSTEMS COMPARED. — RE-
MOVAL OF TREES NEAR THE HOUSE. — CHARACTER OF
TREES, AND THEIR APPLICATION. — THE FORMATION
OF DRESS GROUND. — FLOWER BEDS. — FENCE OF
DRESS GROUND.

HAVING in the two preceding chapters considered the situation and character of the
house, together with the approach to it, under
the first head of improvement—*formation,*—
we come now to treat of the dress ground,
and the scenery beyond it, in the uniting of
which into one harmonious whole lies the art
of improvement, properly so called. As this
discussion will unavoidably embrace both
formation and *removal,* it will be equally as
applicable to the old place as to the new.

Most places, besides the features of the
class to which they belong, have some pecu-
liarities of their own, either as respects the
general expression of the whole, or the circumstances of the parts, as the ground, the
trees, &c.

The eye of taste will carefully observe these varieties, as on the due improvement of them at each place rests, in great measure, the variety of its own character, and its distinction from others of apparently similar features.

Composition in landscape embraces three distinct parts, the *distance*, the *middle distance*, and the *foreground*. The first of these is out of the reach of improvement in itself, but will contribute more or less to the general effect of the scene according to the treatment of those other parts which are under our control. And here it may not be improper to observe, that the very natural pleasure arising from extent of prospect has done much mischief, both in placing the residence and in dictating its accompaniments.

Some years ago I visited a very picturesque spot, upon which an appropriate house was then building. It was a varied knoll covered with full grown wood; the openings here and there carried the eye across a valley adorned with the winding reaches of the Thames to a rich distance beyond. Through one of these openings a distant spire was happily, I should rather say unhappily, seen. A visiter well

acquainted with the geography of the country, to whom the owner of the house pointed
out this fortunate circumstance, informed him
that he might, if he chose it, see from his
lawn seven churches, by removing the trees
that hid them. In an evil hour he listened to
the tempter, and when, some time after, in
passing through the neighbourhood, I called,
in expectation of seeing what had been so
happily begun as successfully completed, I
found the proprietor seated on a bare lawn,
contemplating through a telescope his seven
churches. I have here stated a literal fact,
and, I fear, not a solitary instance, in which
the love of prospect has triumphed over taste,
comfort, and convenience.

An extensive distance is no doubt highly
interesting. The indistinctness of its component parts, and its susceptibility of variety
from every passing cloud, offer that constant
invitation to curiosity which excites the sensa-
tion of cheerfulness in the mind of the beholder. But while

———— " the rude unskilful eye
" Which wild variety with zeal pursues,
" And still is pleased the more, the more it views,"

would lay open the wide extent, —

> " More cautiously will taste its stores reveal
> " Its greatest art is, aptly to conceal;
> " To lead with secret guile the prying sight
> " To where component parts may best unite,
> " And form one beauteous well connected whole,
> " To charm the eye, and captivate the soul." *

I cannot understand Mr. Repton's distinction in the following remark : — " The mind " is astonished and pleased at a very extensive " prospect, but it cannot be interested except " by those objects which strike the eye dis- " tinctly." Nor is it easy to reconcile this observation with another, which occurs a few pages further on, where he says, " By Land- " scape I mean a view capable of being re- " presented in painting. It consists of two, " three, or more, well-marked distances, each " separated from the other by an unseen " space, which the imagination delights to " fill up with fancied beauties, that may not, " perhaps, exist in reality." †

Can the mind be *pleased,* nay, *delighted,* without being *interested?* How different the estimation of an extensive prospect that

---

* Knight's Landscape.
† Sketches and Hints on Landscape Gardening.

suggested the beautiful reflections of the poet! —

> " See on the mountain's southern side,
> " Where the prospect opens wide,
> " Where the evening gilds the tide;
> " How close and small the hedges lie !
> " What streaks of meadows cross the eye !
> " A step methinks may pass the stream,
> " So little distant dangers seem;
> " So we mistake the future's face,
> " Eyed through Hope's deluding glass;
> " As yon summits soft and fair,
> " Clad in colours of the air,
> " Which to those who journey near,
> " Barren, brown, and rough appear;
> " Still we tread the same coarse way,
> " The present's still a cloudy day." *

The middle distance will sometimes be within the influence of immediate improvement, particularly where the domain is extensive. That improvement will depend upon the character of the ground. If it consists of bold swelling forms, the plantations made to vary and enrich those forms may also be managed so as to rise occasionally above the horizon, should it require to be broken; if, on the contrary, the middle distance be of a flatter

* Grongar Hill.

character, the planting should be so effected as to hide a considerable portion of that flatness. In both cases, the plantations should be massive, and their outline varied. If the general occupation of the land be arable, and consequently divided by hedge-rows, a considerable improvement may be effected by planting the corners of some of the fields, so as to unite the angular hedge-row timber into masses of wood.

But, after all, it will most frequently occur, that the principal improvement will be limited to the foreground; and, in all cases, the treatment of that part of the landscape will have the greatest influence upon the whole composition.

Foregrounds, as connected with the subject before us, are of two kinds. One of these may be termed *natural*, as consisting of ground, trees, shrubs, &c., either existing in a natural state, or formed on that model. The other may be called *architectural*, being composed of masonry, as parapets, terraces, flights of steps, &c.

The more I reflect upon the subject, the greater is my astonishment, and the deeper my regret, that the architectural foreground

D

should have fallen a sacrifice to the undistin-
guishing and desolating hand of the modern
system of *improvement.*   Upon what principle of grandeur, of harmony, of propriety,
or comfort, has the exchange been made?

It seems to be universally allowed, that the
habitation of man should be distinct from
that of the cattle that graze around him.   We
see this principle acted upon from the palace
to the cottage; which, with its dwarf wall or
garden pales, broken and enriched with the
simple creepers of honeysuckle, ivy, &c. is an
object pleasing to every eye as well as to that
of the painter.   The variety of material, of
form, and of colour, with the light and shadow
which pervades the whole, are the secret
source of this pleasure.   Strip the cottage of
these accompaniments, and what eye can fail
to regret the destruction?   " What such rustic
" embellishments," says Sir Uvedale Price,
" are to the cottage, terraces, urns, vases,
" statues, and fountains are to the palace
" and palace-like mansion." *   It will be

* I will here remind the reader of my professed object
in these pages, as expressed in the introduction to them ;
viz. to concentrate and render more practically useful the
principles of true taste, diffused through the whole of Sir

obvious that the degree of decoration should vary with the character and consequence of the building which it is to accompany; but let the principle of the architectural foreground be established, and its adaptation to the various circumstances both of cost and situation will be easily adjusted by the eye of taste.

Let us then compare the two systems, with regard to the dress-ground in immediate connection with the dwelling, first, as respects comfort; and, secondly, with reference to propriety, beauty, and picturesque effect.

To seek retirement and protection is natural to man. Hence originated the high walls and close-clipped hedges that bounded the limited gardens of our ancestors; where, on the straight-sheltered walk, the scholar could take his exercise without interruption of his meditation, or relieve his mind by the amusement of his bowling-green, safe from all

---

Uvedale Price's interesting and instructive work: and, in proportion as I may induce the study of that work, in that proportion will be my share in rescuing from destruction all that is worthy to be retained in the old system, and in uniting it with all that is worthy of adoption in the new.

observation. Here, too, the females of the
mansion, secured from every blast and every
intrusive eye, could cultivate the glowing
parterre, which, participating in the same
comfortable shelter, afforded a wreath even
for the brow of Winter.

This sequestered spot usually opened into
the kitchen garden, where broad sunny walks
prolonged the exercise, while a succession of
varied objects imparted a pleasing variety of
sensations to the mind.

These embellishments were amplified with
the extent and consequence of the building
which they accompanied; and terraces, balustrades, steps, &c. increased the variety and
heightened the decoration of the different
mansions, according to their different circumstances and character: still, however, under
whatever modification, comfort and protection characterised the whole.

The modern system throws down the walls,
terraces, steps, and balustrades at " one fell
swoop," and exposes every recess of retirement, every nook of comfort to the blast, and
to the public gaze. The approach invades the
precincts of the garden, which now, in spotty
distinctness, is spread over a space cleared of

every vestige of intricacy and repose; while a sunk fence excludes the cattle from that lawn which is *apparently open* to them, or the flimsy barrier of an iron hurdle is attached to a building whose ivyed battlements have witnessed the lapse of ages.

What compensation, then, does the modern system offer for this destruction of all comfort? Let us consider the question, as we proposed, secondly, as to *propriety, beauty,* and *picturesque effect.*

By *propriety,* I mean that harmony which should invariably exist between the mansion and its accompaniments; and if it be true that external objects affect us by the impression which they make on the senses, and by the reflections which they suggest to the mind, how essential is it that the accompaniments and decorations of the old system should be maintained around the building to which they have been united, perhaps, for centuries! Whoever has visited Powis Castle (as complete in its parts as it is interesting as a whole), may form some idea of the violence that would have been done both to the senses, and the mind, had the *improvement* been there effected which Sir Uvedale Price so feelingly describes, and so justly condemns.

The Beautiful and the Picturesque are so intimately united in the architectural foreground as to be almost inseparable : the Picturesque embraces the leading forms, the angular projections, the abruptnesses, and varieties of the outline; while the Beautiful is traced in the symmetry, regularity, and finishing of the parts.

Let the richness, intricacy, and variety that characterise the old system be contrasted with the best arrangement of ground, the finest verdure, and the most natural disposition of trees and shrubs which modern improvement can effect; and, I conceive, it will be generally allowed that the former will excite an interest, both in the eye and in the mind, beyond any that can arise from the present system; and this in proportion to the magnitude and decorative character of the mansion, as artificial objects require a corresponding accompaniment of art to unite them gradually with the scenery of simple nature.

If, again, we consider the architectural foreground as respects colour, and light, and shadow, the advantage it possesses will be equally obvious. The contrast between the colour of stone and the various tints of vegetation must strike every cultivated eye; while the projections of the parapet, the overhanging coping, the catching lights on the balusters, with the deep recesses between them, broken by the festoons of the various climbing plants, give a play of light and shade highly pleasing; and this architectural arrangement may be more or less accompanied by trees, as the presiding character of the place shall dictate.

The consent to the destruction of all that had cost so much to create, and had imparted so much comfort and enjoyment, could not, in several instances, have been obtained without many struggles between long attachment and the love of novelty, and would be followed by painful though fruitless regret. Sir Uvedale Price's confession might be echoed by all those who had any reverence for antiquity, any feeling for the picturesque.

" I may, perhaps," says Sir Uvedale, " have " spoken more feelingly on this subject from " having done myself what I so condemn in " others—destroyed an old-fashioned garden. " It was not, indeed, in the high style of those

" I have described ; but it had many circum-
" stances of a similar kind and effect : as I
" have long since perceived the advantage
" which I could have made of them, and how
" much I could have added to that effect, —
" how well I could, in parts, have mixed the
" modern style, and have altered and con-
" cealed many of the stiff and glaring forma-
" lities,—I have long regretted its destruction.
" I destroyed it, not from disliking it : on the
" contrary, it was a sacrifice I made, against
" my own sensations, to the prevailing opinion.
" I doomed it and all its embellishments, with
" which I had formed such an early connec-
" tion, to sudden and total destruction."

Some, indeed, would be found alike indifferent to the claim of antiquity and to the suggestions of the Picturesque, — who would
view *change* as *improvement*, and sacrifice every
thing without compunction at the shrine of
novelty.   I was once consulted by the owner
of such a place, who told me, with much self-
gratulation, that I could form no idea of the
labour he had accomplished in the removal
of terraces, sloping banks, &c. so as to reduce
the ground to the state in which I then saw
it — a flat insipid lawn, spotted all over with

distinct shrubs, without connection, without design. The utter insensibility of the owner to any ray of taste relieved me from the painful endeavour to restore a harmony which he had destroyed for ever.

Upon the whole, from a due consideration of the question between the old and new system of landscape gardening, I cannot but think that the former has been sacrificed on account of some tasteless absurdities connected with it, which the early improvers, not being able to separate from it, took the shorter method of destroying the whole, substituting the simplicity of unadorned nature as the accompaniment to the mansion rich in architectural decoration and variety; whereas the architectural foreground, in connection with a shrubbery below it, would lead in an easy gradation to the natural scenery of the park or pasture beyond it.*

And here, perhaps, I may be allowed to express my opinion that the magnificent

---

* Sir Uvedale Price seems to be of this opinion when he says, " Besides the profit arising from total change, a dis-" ciple of Mr. Brown has another motive for recommend-" ing it : he hardly knows where to begin, or what to set " about, till every thing is cleared ; for those objects which " to painters are indications are to him obstructions."

mansion at Burleigh loses half its character and effect from the want of an architectural separation from the park. As it now is, the naked lawn around it, and that only partially mowed, has an unfinished appearance, and excites a regret that some of the original features had not been preserved, or have not been judiciously restored, as the indispensable accompaniments to such a splendid specimen of Elizabethan architecture. There would, doubtless, be some difficulty in the arrangement, from the shape of the ground, and from the living rooms being on a level with the lawn; but I conceive that the richness and embellishment so peculiarly essential to a mansion of that character could be drawn around it with great advantage.

Though the foregoing observations are principally applicable to the buildings of former years, with the hope of preventing the destruction of the architectural accompaniments where they already exist, yet, as I have before stated, I should strongly recommend them (particularly the terrace) to general adoption, regulated by the circumstances of each place, as there are scarcely any situations that might not be improved by the application, while to some it is most essential; as, for instance, when a house stands upon the brow of a steep descent, and where, the soil being unfavourable for the growth of trees, no other foreground can be obtained. Dale Park, in Sussex, is a striking example of such a situation. The house stands on the very brink of a chalky hill, and commands a rich middle distance of park scenery, with an extensive view of the sea in the distance. A bold terrace with its accompaniments, by adding a foreground, would form a beautiful and interesting composition. The decided form of the parapet, with its light and shadow, would, by its contrast, increase the aerial softness of the distance, at the same time that it would hide from the windows the bare unbroken slope of lawn; and, by carrying the eye immediately to the middle ground, leaving the imagination to fill up the intervening space below, would give great apparent extent to the scene.

The architectural foreground is also essential where the ground on one side slopes across the eye with no contrasting line on the other: the terrace wall, in this case, intersecting the sloping line, restores the horizontal

plane upon which a house should stand.   It
will not be necessary that the whole space
between the house and the terrace should be
levelled, where the distance between them is
sufficient to allow of an easy undulation.

I will take the liberty to illustrate my ideas
upon this head from Bromley Hill, so justly
celebrated for the taste it displays.

I conceive that the composition would be
abundantly improved, if, instead of the open
fence showing the awkward slope of the
ground, a horizontal wall and corresponding
line of terrace were carried as far as the first
group of high evergreens, where it might end;
as, from the rapid fall of the ground, the fence
is then lost from the house.   I should return
the wall to the corner of the house, which
would necessarily throw the approach farther
off, and out of sight of the windows, if it were
turned before it ascended the hill ; a point, in
my estimation, of great importance.   These
hints were suggested in a hasty view of a
place where the just and various calls for
admiration left little time for criticism.

The effect of a terrace wall thus applied
may be seen to great advantage at Heanton,

near Okehampton, in Devonshire, from which
the accompanying sketch was made.

Caledon, in Ireland, is an instance of the
effect produced by the architectural foreground. The house stands upon a knoll, the
ground falling every way from it. About
five years ago, I recommended a broad gravel
terrace, with a flight of steps leading to a
second terrace, as a parterre garden. The
good taste of the noble proprietor has added
a third compartment on a still lower level:
and, when I visited the place in October, I
found the myrtles in full bloom upon the
terrace walls, where before no flower could
have endured the exposure of the situation;
while the parapets, vases, &c. form a rich
accompaniment to the mansion, and an appropriate and picturesque foreground to the
scenery beyond it. Perhaps there is no place
where the adoption of the terrace and its
accompaniments has produced a more striking effect than at Clumber. The house on
that side was separated from the park by a
handsome iron fence, almost close to the
windows; from this fence the ground gradually sloped to the water, about a hundred
feet off: that space is now occupied by a

double terrace, the lower one laid out in a parterre garden, and ornamented with vases, fountains, &c.; the whole surrounded by a balustrade wall, with a flight of steps down to the lake. The result fully justifies the undertaking.

Nor is it only as seen *from* the house that the accompaniments of terraces, steps, &c. are productive of that harmony and variety which constitute the Grand, the Beautiful, or the Picturesque effect, according to the situation and character of the building to which they are attached: the extended masonry of the parapet or balustrade, when seen from the approach or the park, gives a base to the superstructure; while the circumstances of steps, vases, &c. mixed with trees and shrubs, produce a richness and variety well calculated to relieve the square mass which characterises the generality of our country residences. Burley on the Hill is a striking example of the good effect of a terrace, as seen from the country around it.

As there are, no doubt, many situations where the terrace cannot be immediately connected with the mansion, it will be necessary to consider the dress grounds under such circumstances, according to their different varieties of character.

Cassiobury stands upon a dead flat; the living rooms upon a level with the lawn: the scenery, as viewed from the window, is principally bounded by the park. A raised terrace would have interfered with this principal feature, and destroyed, in great measure, the cheerfulness of the scene: a broad walk is, therefore, very properly substituted for a terrace.

At Gorhambury, the ground immediately about the house is also flat: but the living rooms, being over a basement story, afford a more varied and extensive view of the park than at Cassiobury. Under these circumstances, I ventured to recommend a sloping bank to be raised, about four feet above the level of the lawn, at a short distance from the house, and parallel with it; and upon this bank there is a broad terrace of nearly four hundred feet in length, the retaining wall of which forms a fence against the deer, while the varied masses of shrubs planted upon it unite it with the flat lawn beneath, and the whole forms a foreground to the scenery beyond. The terrace is connected, by a flight of steps at each end, with the pleasure ground.

As improvement will mainly depend upon the management of trees, including both planting and removal, it may be proper to offer a few hints upon their arrangement under the latter head before we consider the subject of general planting.

With all my partiality for the old system, I would not be understood as deprecating *any* improvement—as recommending *every* thing to be left as we find it. No doubt many points may be yielded to modern comfort and convenience, both in the house and in its accompaniments, without sacrificing the general character and effect.

It may be allowed, perhaps, that shelter rather than taste dictated the deep masses of wood in which some of the mansions under the old system are embedded: in such cases, it is surely lawful to substitute arrangement for quantity—variety for dull uniformity. The operation, however, requires much caution and judgment, especially with trees situated near the house; as an error may be fatal in that situation, which, in a more remote one, might be unobserved, or more easily repaired.

In this, as in every circumstance of improvement, the leading character of the place should guide the hand of the improver.

Though it would be difficult to find any prospect that might not be improved by trees on the foreground, yet they may occasionally be so thick as to render it necessary to break them, both for the improvement of the several groups, and for the general composition.

It is hazardous, on so delicate a subject as this, to give a general prescription : circumstances hardly perceptible to the untutored eye may, to that which has been accustomed to the study of landscape, both in nature and in pictures, be of the greatest moment. A few hints, however, as to what ought *not* to be done, may be safely given ; and I would recommend every proprietor of a place so circumstanced (if he become his own improver) to consult such pictures or prints as are applicable to the case. The " Liber Veritatis " of Claude, and the " Liber Studiorum " of Turner, will afford many examples to the purpose.

The first caution, then, that I would suggest to a person not conversant with the

E

study of landscape, is, not to remove any tree
from the foreground till he has accurately
observed the effect in winter, as well as in
summer. Secondly, *not* to take away a tree
merely upon account of its insignificance, nor
even its ugliness; as the beauty of the group
may be mainly influenced by that very tree.
Thirdly, *not* to seek variety in the group from
the difference of the trees which compose it,
so much as from the general form of the
whole. I would also suggest that round-
headed trees are more picturesque than
pointed ones; though, particularly in connection with buildings, the latter have frequently
a good effect; and, in some cases, are most
essentially useful. There is, I conceive,
scarcely any tree that may not be advantageously used in the various combinations of
form and colour: and, as immediately connected with buildings, I must say that the
Lombardy poplar appears to me to be un-
justly condemned; inasmuch as we have no
tree that so well supplies the place of the
cypress, in contrasting the horizontal lines
of masonry, and giving occasional variety to
the outline of the group. Portman Square
affords an example in point: the horizontal

lines of the houses on each side being broken
and contrasted by the Lombardy poplars in
the plantations; while the plantations themselves derive consequence and variety from
the pointed form and superior height of the
poplars : as, therefore, we cannot command
the cypress of Italian growth, we find the
Lombardy poplar its best representative.

If what has been said upon the advantage
of trees near the house has any foundation
in taste, it follows that the same principle
dictates the planting of trees in similar situations. In doing this, though the immediate
result will bear no comparison with that of
old trees left; yet you have an opportunity of
choice, both as to situation and character of
tree, for future effect, which should be carefully attended to; and then the group may be
thickened with undergrowth, both for shelter
and present appearance. It will be obvious
that these standard trees should be suited to
the soil, and the lawn carried under the group
as soon as can be effected. Sir Henry Stewart's
very ingenious treatise upon the transplanting of trees will be found highly useful in
forming these foregrounds, as it directs the
choice of tree, as well as the mode of re-

moval, so as to produce at once the desired effect.

If a massive foreground of wood, while it excluded an uninteresting country, should at the same time give a sombre effect to the dwelling, I would rather seek to enliven the general effect by decoration, than by laying open a prospect so uninviting; as quantity and richness, even to excess, is preferable to the insipidity of baldness. It is not the least of the advantages of trees near the house, that they create a variety in the scenery as viewed from the different windows, and varying points of the walks. It may, perhaps, sometimes happen, that what would be essentially useful from one window, might interfere with the prospect from another : in such case, the consequence of the windows must decide the question. But more frequently, if properly effected, it will appear that the partial hiding of the scene by foreground trees will not only be a source of variety from the different windows, but that the composition from each will be benefited. It should ever be borne in mind that prospect should not be obtained at the expense of composition. Neither is it from the interior

only that trees near the house are desirable:
they are highly requisite, as accompaniments
to the masonry, when seen from the approach,
and, indeed, from all parts where the house itself is visible.

> " Towers and battlements he sees
> Bosom'd high in tufted trees,"

will apply more or less to every mansion according to its magnitude and character ; and,
in such situations, the Lombardy poplar will
frequently be of essential service.

After all, it will be remembered that *removal*, in common with every measure connected with improvement, must rest mainly
upon the local circumstances of the place, and,
consequently, will only admit of general suggestions by which the proprietor may be
awakened to the subject; so far, at least, as
not to destroy the character of his place by
substituting the baldness of the ,modern
system for the rich formality of the old
school. The hand of taste will cautiously
withdraw the veil which separates them, and,
by degrees, admit the surrounding scenery,
without destroying the shelter and partial
seclusion so essential to the mansion we have

been considering : cheerfulness, rather than gaiety, should be the proposed result.

The following remarks upon the treatment of the dress ground may occasionally apply to the old mansion ; but it is to assist in the formation of a new place that they are principally directed : —

> —— " To deck the shapely knoll
> That, softly swell'd and gaily dress'd, appears
> A flow'ry island, from the dark green lawn
> Emerging, must be deem'd a labour-due
> To no mean hand, and asks the touch of taste." *

The dress ground immediately connected with the house should be considered as the foreground of the picture, which the whole scene, taken together, presents to the eye, and should be treated as such.   The groups, and single trees upon it, should be planted with reference to the scenery beyond, so as to lead the eye into the remote parts of the picture ; excluding, as far as may be, whatever might injure the general composition.

In the formation, then, of the dress ground, I should recommend the making a slight sketch from the leading points of view,

* Cowper.

(usually the windows of the library or drawing-room,) of the general scene as it *exists;*
and then add to your sketch such groups of
trees and shrubs, and such detached trees, as
would hide the less interesting parts of the
landscape, and, by breaking the uniformity of
other parts, produce that connection so essential to composition. In forming such
groups, particularly of larger trees, it should
be well considered, whether a massive or a
lighter group is requisite; whether the most
distant scenery is to be caught through the
stems of these trees, or to be altogether excluded by them. I would plant all the larger
features with this reference to the general
scene, before proceeding to the lesser embellishments of the lawn, as flower-beds, &c.
which should be formed with reference to
those features.

The groups of larger trees will usually be
accompanied by shrubs of various size and
character, to connect them with the lawn:
rhododendrons, savine, and other of the
pendent evergreens, are very useful for such
purpose, when the turf, being carried under
them, leaves no cutting line of border.
Shrubs, in my opinion, should not be accom-

panied, in the same bed, by such flowers as
require digging; the line of border above
mentioned destroying that repose and that
variety of form which ought to characterise
the former. In a lawn of small dimensions,
the loosing of the turf under the shrubs is of
the utmost importance, as it gives an appearance of extent to its limited proportion.
Pæonies, roses, hollyhocks, and other flowers
that are of sufficient height or size to mingle
with the shrubs, may be fairly united with
them, if it can be effected without showing
the mould. In the first formation of these
plantations of shrubs, the borders must be
dug, and, for a time, kept so; but every opportunity should be taken to break the edgy
line, till it can be finally obliterated : to help
this end, even in the first instance, periwinkle,
St. John's wort, and other ground creepers,
may be planted with the shrubs; and, by
uniting them with the lawn, will tend to di-
minish the hard line of the border : a thing
that cannot be too strongly insisted upon, as
essential to continuity and repose.

It is impossible to lay down rules that may
regulate the size, situation, or character of
these plantations of shrubs, which will depend

upon the shape, character, and extent of the grounds they are to embellish : some hints, however, as in other cases, may be suggested, to direct the unskilful hand in an operation of no small importance.

And first, as before observed, the plantations should be marked out from the principal point of view, so as to agree with the general scene. The size of each mass will depend partly upon what is to be excluded or broken in the remote landscape, and partly upon the character and size of the ground. If the situation to be planted be of small dimensions, one mass of tolerable size may be better than dividing it : but, if the ground admit of it, a variety of masses is preferable, as producing more intricacy and greater appearance of extent. In this case, the masses of shrubs will be so disposed as to show portions of lawn intersecting them in glades of different size and form. The general inclination of these masses of shrubs should tend, though in different degrees, towards the most interesting part of the scene, either within or without the dress ground, as circumstances may be. A horizontal line of plantation can rarely have a good effect.

Though there will be a variety in the forms
of the plantations, there should be a general
harmony of outline between them when they
approach each other; the more swelling part
of one opposing itself to the recess of the
other. The nearer masses should generally
be of lower material than the more remote,
that the one may occasionally be seen over
the other. Where the dress ground is of
such dimensions as to allow these masses
upon a large scale, the variety of their re-
spective forms should be boldly marked, as,
in the course of a few years, they ought to be
broken. A few of the choicest plants will
then occupy the space that present effect
requires to be filled up with common material. For this purpose, care should be taken,
in the first instance, to dispose of the choice
plants before the mass is filled up, so
that the former shall hereafter stand where
they ought; whereas, for want of this precaution, a cedar of Lebanon may, when grown
up, destroy the composition which, had it
been rightly placed, it would have materially
improved. For places where a cedar of
Lebanon, or any of the larger firs, might be
thought too big, the Virginia cedar or the

hemlock spruce is well adapted : the latter is, in my opinion, the most beautiful of
that class of evergreens ; it likes shade and
a moist ground, though I have seen it flourish
in drier and more exposed situations.

When I said that these masses of shrubs, &c.
should be marked out from the principal
point of view, I did not mean that they need
not be studied from any other point : on the
contrary, it is essentially necessary that they
be examined from every situation from which
they are to be seen, that no beauty, as far as
can be avoided, may be lost through inattention.

In order to assist the arrangement of these
plantations, I should recommend forming
them on the ground, with branches of various
lengths, with the leaves on, which gives a far
better idea of the intended effect than can be
given by stakes : the branches, being laid on
the ground, can be turned in any direction,
till the best forms are obtained, which may
then be marked on the turf with the edging-
iron: Larger branches, stuck in the ground,
will direct the placing of trees with the same
advantage over a mere stake. Good hints
for such planting of a lawn may be found on

any common, where furze, broom, &c. furnish
endless varieties of form and grouping. Having disposed the masses of trees, shrubs, &c.
with reference to the general effect of the
whole scene, we come now to the finishing
touches of decoration — flowers.

From the general love of flowers, and their
increasing varieties, we frequently see the
breadth and repose of the lawn sacrificed to
them. In a flower-garden, properly so called,
they hold undivided sway, and are at liberty
to cover the whole surface, and to assume
every variety of form that fancy may dictate; but, when flower-beds are component
parts of the dress ground we have been considering, they must be amenable to the rules
of composition, otherwise they injure the
scenery they are intended to adorn. Beautiful examples of the former arrangement (the
flower-garden) will be found at Cassiobury
and at Redleaf; the combination requisite
to the latter will be found in equal perfection
at Danesfield.

The disposition of flower-beds will vary
with the character of the house, and the extent and circumstances of the ground about it.
At the manorial building, where the straight

walks and the appropriate accompaniments
are retained, the flower-beds should, in my
opinion, be characterised by the same pre-
cision and regularity. I have treated them
upon that principle at Somerhill, one of the
finest specimens of the Elizabethan mansion
with which I am acquainted.* As, however,
beds of this description, being necessarily
filled with flowers of low growth, have rather
a flat and tame appearance, their effect will
be greatly improved by a border, which will
elevate them above the lawn, and, by producing a variety of light and shadow, will give
richness and variety to the mass. The border
may be made of wood or iron, painted to resemble stone, which will unite them more
harmoniously with the masonry of the house,
terrace, walls, &c., at the same time that it will
relieve them from the lawn better than any
other colour. The height of the border will
depend upon the size of the beds : for those
of moderate size, about six inches will be
sufficient. When of larger dimensions, a foot

* It is hoped that such references to different places
will be taken as they are intended, viz., to give an opportunity of comparing the principles laid down with their
actual effect.

is not too high.   The effect of flower-beds so
constructed may be seen in the garden of
Lambeth Palace.

Where the character and decoration of the
mansion will warrant it, these borders might
be made highly ornamental, and might, I
conceive, be cast in iron at a moderate expense.   The effect, even in the simple style,
will be improved by the introduction of
vases, flower-stands, and orange trees, or
other shrubs, in handsome tubs: the flower-
stands should not be of rustic character, but
of regular form and exact finishing.   Wood
or iron is prefesable to stone, as less exposed
to injury from the roller.

In what may be termed a free disposition
of flower-beds, the first care should be to
avoid the spottiness which must result from
putting a bed wherever room can be found
for it: on the contrary, the beds should be
treated upon the same plan of composition
that arranged the shrubs they are to accompany.   The glades of lawn that have been
created by the foregoing operation must not
be destroyed by scattered beds of flowers
crossing them in all directions; though occasionally a bed will be introduced to break

the continuity of the line of shrubs, and relieve, by brilliancy of colour, their more
sober tone. As breadth, however, equally
with connection, is essential to composition,
the beds, in general, should be brought together in masses, leaving lesser glades among
them: and these glades, again, should be
broken by a single plant or basket, taking
care never to place such interruption *midway*
between the sides of the glade. The masses
themselves will be lightened by a detached
bed or two of a lesser size. There is no objection to the occasional introduction of a
regular form in the flower-beds; though, for
the most part, the easy curving lines will
unite better with each other. Baskets and
picturesque stands are also useful to relieve
the flat surface of the masses, if they are not
too profusely introduced. It may here be
observed, that, though *basket-like* forms may
be applied to beds of a large size, the *handle*
should not be added to any one longer than
appears capable of being lifted, as the want of
proportion is too glaring: and the handle
itself cannot be enriched so as to be well
united with the contents of the basket.

Gravel walks being necessary to the enjoyment of the scenery we have been considering, it may be useful to offer a few observations upon this part of our subject. The line of walk should, I conceive, be regulated by the size and circumstances of the place. And, first, of whatever extent the grounds may be, I would never carry the walk round the boundary; nothing, as I have before observed, is, to my feeling, so insipid as a long-continued sweep : and the hanging perpetually on the boundary, by betraying the real dimensions of the place, destroys all idea of extent as effectually as it does that of variety. Whoever has seen the pleasure-ground at Caversham (laid out by Brown), cannot but perceive what an improvement it would be to wind the walk amongst the noble trees and rich masses of shrubs, which now trails its monotonous course by the side of the sunk fence.

There are situations where the walk cannot, in any direction, be carried round the dress ground, without manifest injury to the general effect; as, when the lawn in front of the house is flat and of small extent ; and which, for the sake of the scenery beyond, must be kept open and unbroken by plant-

ing for a considerable space. A walk, under
such circumstances, would destroy the repose
of the lawn, and, at the same time, narrow its
extent by placing it between two lines of
gravel, as there must be a walk close to the
house. In such a case, the house-walk (as we
will call it) may be taken on either side, as
may best suit, through the closer plantations,
and may be returned into itself, out of sight
of the windows, leaving the lawn to be paced
in any direction that the variety and richness
of its glowing decorations may invite.

The above remark was suggested to my mind
by a perfect example of its propriety, and
which is at this moment before me. You
step from a colonnade filled with the gayest
flowers, upon the walk, that, passing a conservatory, leads you under a canopy of overhanging trees, — a short but beautiful circuit,
which returns you to the lawn, sparkling with
flowers, the arrangement of which I have
already noticed. This lawn falls in a varied
slope, till it is lost in the recesses of a woody
skreen, which shelters it at once from the
northern blast and from the obtrusive gaze
of the approach; while far below are seen
the remains of Medmenham Abbey, backed

by magnificent elms standing upon a winding
reach of the Thames, enlivened by the circumstances of a ferry boat, and other craft
passing up and down the river. A group of
horse chestnut, under the branches of which
the Abbey is seen, forms the foreground, and
finishes the picture. To those who are acquainted with this beautiful, and, I may add,
unique spot, I need not name Danesfield. In
laying out the walks, care should be taken to
keep them as much as possible out of sight
of the windows, and also of each other; as
seeing one walk from another gives an idea
of limitation. Where an occasional portion
of walk thus intrudes, it may be hid by raising
the turf; but this should be effected by a
gentle slope. A terrace walk, even at the
shortest distance, will not offend, either from
the windows or the walk under them, as it is,
if I may so speak, a limitation of choice, not
of necessity. I am not partial to *berceau*
walks: they belong, at any rate, rather to
the flower-garden than to the shrubbery, of
which we are now speaking: they may, however, be occasionally well applied; as in
leading to a seat or ornamental building in
character with them; or, still more appropri-

ately, in masking some boundary fence which cannot easily be avoided: but long walks of this description, intersecting the open lawn, I think, are sadly misplaced.

The width of a walk must in great measure be determined by circumstances. If it be of such extent as to afford the means for exercise, it should admit of three persons walking abreast; as otherwise one is thrown out: this accommodation cannot be obtained under a width of six feet, but I think seven better. When the walk is of too limited extent to be used for exercise, its width should be in proportion to its length, and to the scale of the grounds. It is desirable, however, that the walk near the house, where it can be done, should be of sufficient length and width to supply the want of a terrace, where the latter cannot be obtained. If this walk can be made a straight one, it will answer the purpose better than if curved, and its width may be more easily accommodated to that of the narrower walk continuing from it than can be effected on the latter form. When one walk breaks off from another, it should be at a right angle, thus avoiding a sharp point of lawn between them, which it is difficult to break by any shrub or other decoration.

Having now planted the dress ground, given it the last touches of decorative finishing, and carried the walks through its varying scenery, it becomes necessary to protect it from the incursions of the cattle that graze the pasture from which it has been taken.

The observations I have ventured to make a few pages back, express my opinion upon the absolute *necessity* in many cases, and the great utility in many more, of an architectural fence between the dress lawn and the country beyond it. As, however, there are various situations to which those observations will not apply, it becomes necessary to enter more largely into the subject of fences.

That the fence should vary with the character of the place, might have been expected to be generally allowed; experience, however, proves the contrary: we see the sunk fence, or the iron hurdle, applied indiscriminately to the mansion of two centuries' standing, and to the erection of yesterday; to the castle, and to the cottage. There are, it must be allowed, many degrees of finishing in the latter; from the common hurdle to what is called an invisible fence; the last, the most objectionable, upon the principle I wish to recommend.

I cannot but think that (with the exception of Sir Uvedale Price) the different writers upon the improvement of scenery as connected with residences have, as far as I am acquainted with them, altogether mistaken the question of a separating fence. They think it essential that no visible interruption should exist between the smooth and decorated lawn and the scenery, of whatever description, beyond it. To effect this junction they have recourse, as the happiest expedient, to a sunk fence; yet, fearful of detection, they recommend various modes of hiding this *invisible* fence; in effecting which, they are likely to raise a far more objectionable line of separation than the rudest fence would be.

The author of Observations on Modern Gardening, from whom better taste might have been expected, entangles himself on this subject in the following observations:—

" The use of a fosse," says this writer, " is merely to provide a fence without ob-" structing the view. To blend the garden " with the country is no part of the idea;

" the cattle, the objects, the culture, without
" the sunk fence, are discordant to all within,
" and keep the division. A fosse may open
" the most polished lawn to a corn-field,
" a road, or a common, though they mark
" the very point of separation. It may be
" made on purpose to show objects which
" cannot or ought not to be in the garden ;
" as a church, or a mill, a neighbouring gen-
" tleman's seat, a town, or a village, and yet
" no consciousness of the existence can re-
" concile us to this division. The most
" obvious disguise is to keep the hither
" above the further bank all the way; so
" that the latter may not be seen at a com-
" petent distance : but this alone is not always
" sufficient, for a division appears, if an uni-
" formly continued line, however faint, be
" discernible ; that line, therefore, must be
" broken : low but extended hillocks may
" sometimes interrupt it ; or the shape on
" one side may be continued across the sunk
" fence on the other ; as when the ground
" sinks in the field, by beginning the declivity
" in the garden. Trees, too, without, con-
" nected with those within, and seeming part
" of a clump, or a grove, will frequently

" obliterate every trace of an interruption.
" By such or other means the line may be, and
" should be, hid or disguised; not for the
" purpose of deception (when all is done we
" are seldom deceived), but to preserve the
" continued surface entire. If, where no union
" is intended, a line of separation is disagree-
" able, it must be disgusting when it breaks
" the connection between the several parts of
" the same piece of ground. That connec-
" tion depends on the junction of each part
" to those about it, and on the relation of
" every part to the whole. To complete the
" former, such shapes should be contiguous
" as most readily unite; and the actual di-
" vision between them should be anxiously
" concealed. If a swell descends upon a
" level, if a hollow sinks from it, the level
" is an abrupt termination, and a little rim
" marks it distinctly. To cover a short
" sweep at the foot of a swell, a small rotun-
" dity at the entrance of a hollow must be
" interposed. In every instance, when ground
" changes its direction, there is a point where
" the change is effected, and that point should
" never appear; some other shapes, uniting
" easily with both extremes, must be thrown

" in to conceal it. But there must be no
" uniformity even in these connections: if
" the same sweep be carried all round the
" bottom of a swell, the same rotundity all
" round the top of a hollow, though the
" junction be perfect, yet the art by which it
" is made is apparent; and art must never
" appear. The manner of concealing the
" separation should itself be disguised; and
" different degrees of cavity or rotundity,
" different shapes and dimensions to the
" little parts, thus distinguished by degrees;
" and those parts breaking, in one place more,
" in another less, into the principal forms
" which are to be united, produce that va-
" riety with which all nature abounds, and
" without which ground cannot be natural."*

Allowing, for the present, the justice of
the theory here laid down, what possible
chance is there of its being executed so as to
effect the purpose intended? The management of ground requires the greatest skill,
even when the scale of operation is of considerable extent and breadth of character.
With what hopes of success then can these

* Observations on Modern Gardening, p. 8.

" *low but extended hillocks*" be attempted? To whom could be committed the delicate oper-ation of a *connection that depends on the junc-tion of each part to the whole ?* To what hand could be intrusted the *manner of concealing the separation, which should itself be disguised* —*the different degrees of cavity or rotundity;* —*the different shapes and dimensions to the little parts thus* distinguished *by degrees,* &c. &c.? If this intricate operation were to be effected under direction that could afford any pro-spect of success, the expense would outweigh the advantage proposed : if it were committed to other direction, the attempt would be worse than abortive. But, added to all this, these low but extended hillocks, &c. are, many of them, if not all, calculated for a single point of view, as change of place would materially derange the effect intended by them.

Let us, however, see how far the author's observations are founded upon the principles of taste. He tells us the use of a fosse is merely to provide a fence without obstructing a view : he here takes it for granted that *no* view is to be obstructed; his prescription is of universal application : a corn-field, a road, or a common; a neighbouring gentleman's seat, a mill, a town, or a village, are equally objects to be shown; at the same time he admits these objects to be *discordant to all within.*

It is in such circumstances as this that the study of landscape in pictures, as well as in nature, appears to be essential in qualifying an improver for his profession. Such study would have shown the author of the above observations, that *composition*, not *view* only, is the object to be aimed at. How, for instance, would the nicest concealment of the fosse ever reconcile to the eye of a landscape painter such ground as we were considering a few pages back, where one uniform slope passes across the eye, with no contrasting form to balance it?

Mr. Mason, in his Essay on Design in Gardening, treats the subject of fences as follows:—

" *Uniting* the scenery in landscape is the
" chief purpose of sunk fences. Wherefore
" they should be perfectly concealed them-
" selves, that we may not discover insuf-
" ficiency in the execution: neither should
" unnatural swells of ground be made use of

" in order to conceal them ; for thus the very
" purpose of uniting must be defeated.

   " The author of Observations on Modern
" Gardening enters (p. 8.) on this subject of
" fosses ; but in so superficial a manner as
" plainly shows, either that he was but little
" acquainted with the principle of their ap-
" plication, or did not choose to encounter all
" the difficulties of reducing this principle to
" practice. But the poet, in the second book
" of the English Garden, goes fairly into the
" subject of sunk fences, and describes the
" best that can be made, both for internal
" and external deception. He acknowledges,
" indeed, that such contrivances are

—— ' defective still,
' Though hid with happiest art.'

" Yet one consequential defect he certainly
" palliates. To say that the scythe on one
" side, and the cattle on the other, ' create a
" ' kindred verdure,' is more poetical than
" exact. The cattle always leave something
" which the scythe does not leave, and suf-
" ficient to mark the line of separation to a
" common eye. This defect, indeed, may
" sometimes be easily cured, by only using

" the scythe a little way on the outside; for by
" this method the extremity of the scythe's
" dominion may be made so conspicuous as to
" preclude any suspicion of deception *there*,
" and mere change of cultivation will not
" alone spoil harmony of landscape.   Where
" the junction is easy, we still admit

> ' The useful arable and waving corn,
> ' With soft turf border'd.'   *Shipley.*

" But sunk fences, wherever visible, are so
" manifestly artificial, that a good designer
" will take great pains to secure their perfect
" concealment, and rather have recourse to
" any other practicable mode of harmonizing
" landscape.

" One other method, by which we are to
" annihilate the view even of an upright rail-
" ing, is given us by the same poet. His
" way of doing it is with an invisible colour;
" and an admirable expedient it would be, if
" the theory would hold in practice; which,
" I apprehend, it will not. The receipt in
" the poem is quite enigmatical—not, how-
" ever, inexplicable as to the materials; but
" the proportionable quantities of each are
" left very much at large; and I never could

" meet. with any mixture of them that per-
" fectly answered the purpose. The chief
" use of such colour would, in my idea, be
" hiding gates between enclosures, where they
" could not so well be hidden by any other
" means; for as it is impossible the fallacy
" should succeed within a moderate distance
" from the eye, a length of such fences can
" never be eligible. The poet very justly
" observes, in his postscript, that the con-
" cealment of fences is a matter of great
" difficulty both to design and to execute:
" for which reason it may not be amiss to
" dwell a little longer on the subject. And
" here I repeat, that *harmonizing a landscape*
" is always the point to be aimed at. Uniting
" different enclosures, and giving an air of
" unlimited extent to the premises, may be
" consequential incidents, but should never
" be considered as a principle to work by.
" As far as vision is concerned, taste, in
" Shenstone's language,

' Appropriates all we see.'

" But (without any reference to actual pro-
" perty) a narrow line of partition is of
" itself a disagreeable object; and wherever it

" obtrudes upon the sight in such a form,
" necessarily destroys harmony of landscape.
" A place, however, must be very destitute
" of inequality of ground, not to admit a
" change in the nature of the narrow line by
" low plantations adjoined to it, without ob-
" structing the view above it. There are
" shrubs of every stature (down to the creep-
" ing perrywincle) proper for this purpose
" within a garden, and there are hollies and
" thorns for pastures."

Can any thing be more superficial than
these observations? and yet their author
applies that term to the elaborate discussion
we have just been considering, as taken from
the Observations on Modern Gardening.
Mr. Mason's ideas upon the subject, I think,
are not to be ascertained from the above
extract; the only use of which is, that it
affords a proof (in addition to many others) of
the inutility of suggestions not founded upon
some principle.

The acknowledgment of the author of the
English Garden, after all his investigation of
the sunk fence, that it is

―――― " defective still,
" Though hid with happiest art,"

renders it unnecessary to go through the subject with him; and I fear his receipt for annihilating an upright railing is equally defective.

Mr. Repton, treating of fences to the dress ground, says,—" After various attempts to " remedy these defects, I have at length " boldly had recourse to artificial manage- " ment, by raising the ground near the house " about three feet, and by supporting it with " a wall of the same material as the house. " In addition to this, an iron rail on the top, " only three feet high, becomes a sufficient " fence, and forms a sort of terrace in front " of the house, making an avowed separation " between grass kept by the scythe, and the " park fed by deer, or other cattle."

In many instances, this raising of the ground must have a bad effect; nor, I fear, would the iron rail make " an avowed sepa- " ration" between the dress ground and the pasture.

In point of expense nothing is saved, as the supporting wall is to be of the same cha- racter as the house; and consequently would serve for the dwarf wall I recommend.

But whence this horror of a fence, which

good sense—a constituent part of good taste
—prescribes? If it be contrary to good sense
to admit the cattle on the dressed lawn, it is,
I conceive, equally contrary, to let it appear
they are admitted.   The observations I have
ventured to make a few pages back, express
my opinion upon the absolute necessity in
many places, and the great utility in many
more, of an obvious and solid fence between
the dress ground and the country beyond it.
I would not, however, be understood as pre-
scribing a wall for the appropriate fence, in
all cases. To prevent such misapprehension,
it may be necessary to enter more largely
into the subject.

And first, there are places where no separating fence is visible, either on the dress
side or on that of the approach; as at Wilton, and at the Priory near Stanmore. The
former, being entered by an enclosed court-
yard, leaves such extensive grounds on the
dress side, that the fence is lost amongst the
masses of trees and shrubs with which it is
adorned. The entrance to the latter, though
open to the park, is also completely excluded
from the dress side of the house: and, in a
pleasure ground of fifty acres, melting with

natural gradation into the scenery beyond
no distinct line of fence is seen or required.
So also at Clumber, the only visible fence
is the terrace wall, which extends from the
house to the river, the latter then becoming
the separating line. When, however, a fence
is attached to, or seen from, a house of the
old character, I hold it essential that such
fence be of masonry, even where circumstances do not admit of a terrace. I was
much pleased to find my idea realized at
Cassiobury by a corresponding feeling; and
I could not name a place where the effect
is more completely illustrated.

Nor is it only in connection with houses
of the old school that I should recommend a
dwarf wall as the separating fence. In all
houses which approach to the consequence
of a mansion, if circumstances permit, I
should wish its adoption: more particularly
where it is essential that the uniformly in-
clined line of the scenery be interrupted, as
described in the observation on Bromley
Hill, a transparent fence will not restore
the horizontal plane so necessary to the
composition, as the sloping ground beyond
will be seen through it.

We have already seen that some places are so circumstanced as to require no fence visible from the windows: there are, also, others, though of smaller dimensions, where, from the inequality of the dress ground, the fence will be lost among the shrubs in the bottom. Neither will a wall be applicable where the lawn falls laterally as seen from the windows, and cannot be planted out without injury to the view. Danesfield is an example in point, where the lawn, having passed the windows in a horizontal direction, falls rapidly down till it is lost in a wood below: here, however desirable a dwarf wall might be on the horizontal plane of the lawn, any attempt to plant out its junction with the fence on the descending line would be highly detrimental.

In places where the dwarf wall is applicable, its situation with regard to the house will vary according to the different circumstances of each place. When the living rooms are on, or nearly on, a level with the lawn, and where the scenery is rather flat than elevated, the wall should not be far from the house, as it would shut out too much of the view. And here again, Cassiobury furnishes an apt example; the space enclosed by the

wall before the windows being small, while
the pleasure ground stretches into quantity
in another direction. The wall also is, with
great judgment, varied to agree with the
varieties of projection and recess of the mansion, as one straight line of such length
standing on the flat surface of the lawn would
be very insipid; whereas the different rectangular breaks, with the light and shadow
resulting from them, give a variety and richness highly pleasing.

The height of the wall will be governed in
like manner by its relation to the circumstances
of the house. The one, which we have been
considering, is two feet six inches high; a
greater height would have interfered with
the scenery. Where such interference is not
apprehended, I think three feet a better
height, as seen from within; and by sloping
the ground without, so as to get four feet,
you have a sufficient fence except against deer;
where that is necessary, a slight iron wire
addition on the top of the wall will answer
the purpose, and be scarcely visible from the
windows. Previous to building the wall, I
should recommend trying the effect both of
its height and situation, by throwing a garden

mat over a pole a few yards long, which may be shifted till the best situation, &c. is ascertained. I have met, occasionally, with places, where to fill up an existing sunk fence would be very expensive : in that case, I would erect a wall on the inside of it, so as to remove all idea of an invisible fence ; a skirting wall of a foot or eighteen inches high will effect this, if circumstances, either of cost or situation, forbid a higher. Where stone is not easily procured, the wall may be built of brick, and splashed to resemble stone, which is the case with the wall at Cassiobury, as it is, indeed, with the house itself. Where the walk accompanies the line of wall, the effect, I think, is better, when it is unbroken by any creepers; but, where there is a space between them to be filled with flowers, then the festoons of creepers give an appropriate and beautiful variety to the masonry.

When a terrace is formed to a building of a regular front, it is desirable, where no obvious impediment prevents, that the extent of the wall should be at equal distance from each end of the house ; but, where an adequate impediment interferes, the irregularity of extent is satisfactorily accounted for ; but

I think the extended side should, in that case, be of such length as to preclude all idea of agreement with the other.

To sum up, in few words, my ideas upon the subject of fences:—I hold it imperious that a manorial house, either of ancient or modern date, should be separated from the pasture by a wall. I think it agreeable to good taste, that a Grecian, Italian, or any other pile of sufficient character or magnitude, should also be thus accompanied. In cases were this accompaniment is not requisite, or cannot well be applied, I prefer a more solid fence to a flimsy one; and a sunk fence I hold to be totally irreconcileable to a shadow of taste. It will be remembered, I am speaking of the division between the dress ground and the pasture beyond it. To more remote situations, where it may be desirable to remove a hedge, and yet retain the division of the grounds, the least visible separating line will be the best adapted to the purpose, and a sunk fence may be as good as any other. It will also be remembered, that I am recommending a wall only where the dress lawn is seen in conjunction with the pasture.

Before we quit this subject, it may be useful to notice an arrangement of Mr. Brown's, as destructive of cheerfulness as it is destitute of taste, viz. the enclosing by a sunk fence a large portion of ground beyond the dress lawn (from which it is separated by the same expedient), and planting both the sides, while the remote front is left open to admit the distant view. Within this sunk fence, but on the outside of the plantation, a monotonous walk leads you round the confines of this cheerless patch of coarse grass, which, being neither ornamented nor fed, is intended as an apparent continuation of the velvet turf surrounding the mansion. A stronger instance of mistaken theory and practice in the art of gardening, I think, is scarcely to be met with. I trust this arrangement is improved at Woolterton, in Norfolk, and at Kirklinton, near Woodstock, by substituting a terrace, and carrying the walk in a varied line through the plantation, now grown into fine trees, and by the planting of groups of ornamental shrubs in the enclosure at the one place, and at the other, by throwing it open to the sheep, according to the different circumstances of each.

CHAP. IV.

PLANTING. — ERRORS COMMITTED. — IRREGULAR FORM IN OPPOSITION TO OVALS AND CIRCLES. — CONTROVERSY BETWEEN SIR UVEDALE PRICE AND MR. REPTON.

FROM the dress ground we pass to the scenery beyond it. As the beauty and character of this part of the picture will depend (as far as art can assist it) chiefly upon planting, some general hints may be given on that head, for conducting it so as to show and improve such varieties of ground as the place may possess, though it will not be possible to give a plan that shall be applicable to all cases.

One rule, indeed, may be universally laid down — never to plant a belt.

In planting, the first care should be to connect the different plantations under one general intention ; not to scatter them in detached spots, as it were at random, without any purpose of uniting them in composition.

How frequently do we see undulations of ground, which might have been infinitely varied by judicious planting, utterly deformed

by a cap of fir or larch placed on every swell.
Whereas, had the plantation on one knoll
extended into the hollow, it would have more
strongly marked the depth between it and
the corresponding swell on the other side of
the valley; on which swell a looser plantation
might flow half way down, connecting it again,
by a straggling group or two, with some other
mass of wood.

No. 72. of the Quarterly Review incul-
cates this lesson with great force and taste:
the passage will be found worthy of attentive
perusal.

" The improver ought to be governed by
" the natural features of the ground, in choos-
" ing the shape of his plantations, as well as
" in selecting the species of ground to be
" planted. A surface of ground undulating
" into eminences and hollows forms, to a
" person who delights in such a task, per-
" haps, the most agreeable subject on which
" the mind of the improver can be engaged.
" He must take care, in this case, to avoid
" the fatal yet frequent error, of adopting
" the boundaries of his plantation from the
" surveyor's plan of the estate, not from the
" ground itself. He must recollect, that the

" former is a flat surface, conveying, after the
" draughtsman has done his *best*, but a very
" imperfect idea of the actual face of the
" country, and can, therefore, guide him but
" imperfectly in selecting the ground proper
" for his purpose. And again, the man of
" taste will be equally desirous that the boun-
" daries of his plantations should follow
" the lines designed by nature, which are
" always easy and undulating, or bold, pro-
" minent, and elevated, but never either stiff
" or formal. In this manner the future woods
" will advance and recede from the eye ac-
" cording to, and along with, the sweep of
" the hills and banks which support them,
" thus occupying precisely the place in the
" landscape where Nature's own hand would
" have planted them. The projector will
" rejoice the more in this allocation, that in
" many instances it will enable him to con-
" ceal the boundaries of his plantations ; an
" object which, in point of taste, is almost
" always desirable."

In forming plantations, either of larger or
smaller dimensions, I should strongly recommend, in agreement with the above quotation,
a varied form instead of the lengthened straight

line or gentle curve of the former, and the oval or circular figure of the latter, which have so generally prevailed. The beauty of a wood depends mainly on the beauty of its outline: and that outline requires a variety, which can never be found in an insipid sweep, but which arises from the contrast of projection and recess; remembering that small variations will not correct the insipidity, and that the effect will be good in proportion to the boldness of the contrast. These recesses, again, should vary from each other both in size and character; but, in all, an angular abruptness should be preferred to a smoother form: and, above all, the connection of the several parts into one harmonious whole should be ever kept in view. The nature of the lower growths, as thorn, holly, &c., is essentially useful in producing these varieties of character, by giving density to some parts, whilst others will admit the eye through the boles of the more open grove into the interior of the wood; thus producing that variety and intricacy which a natural wood seldom fails to exhibit.

It is very necessary to notice an error too prevalent in forming large masses of wood; I mean planting the whole surface, and trusting to future removal for producing that variety acknowledged as essential to the intended effect. By this mistaken plan, those undulations of ground, upon which the beauty of the plantation will mainly depend, are buried, and never can be restored with any thing like original character and effect: add to which, the future outline will be described by trees more or less deformed by their interior situation, and deprived of that drapery, if we may so term it, which should break the swelling line, and overhang the receding hollow.

I remember passing by a wood belonging to the Duke of Buccleugh, I think in the neighbourhood of Ecclefeckin, which, from having been partially burnt, offered a perfect model for the mode of planting above recommended.

What has been said with regard to the outline of a wood, will apply equally to a clump, as it is called; preserving a due proportion according to its extent: indeed, the bad outline of a clump is, perhaps, more offensive than that of a wood, as the massiveness of the latter, in some measure, atones for the poverty of its outline, while the beauty of the clump depends almost entirely upon its form.

It is difficult to conceive how any person, conversant with the varieties and combinations of nature (which every improver should be), could ever stumble upon so monotonous a form as an oval or a circular group of trees. If variety and intricacy are essential to picturesque effect, what of either is to be found in these figures, whether in their youth, or when released from their enclosure?   Let any one, conversant with the subject, examine an oval or circular plantation of any age, and try how many trees he can preserve in the endeavour to give it any resemblance to a natural group: nearly all within are poles; and so many must be removed from the circular line, ere that line can be at all obliterated, as will leave at last a very small proportion of the number originally planted. I speak, I may say, from painful experience; having frequently been under the necessity of inverting the principle of decimation by the removal of nine out of ten, to obtain even a tolerable combination.

On the other hand, the irregular form

offers, I conceive, every facility for future improvement; as by separating altogether some of its projecting points, you obtain detached groups of varied size and character, and yet in connection with the larger mass: added to which, the groups thus separated will consist of well furnished trees, from their having been exposed to the air and sun since they were first planted.

As I cannot but think it self-evident, that the *future* effect of the irregular must be preferable to any that can be obtained from a regular clump, so I conceive its *present* appearance to be abundantly better. View the regular form on which side you will, it is a dense mass of unvarying shape and surface: whilst the irregular is a continued variety of form as you move round it; and, from its angular projections and recesses, affords that light and shade which is sought in vain from the uniform curve. It is necessary that groups or clumps should be of different size as well as of different form, as similarity of appearance marks them as works of art; one great objection to the regularity of form in the oval or the circle.

The upper sketch is an exact representation

of the Park at Sledmere, in Yorkshire : the oval plantations, though of about a thousand feet in circumference, do not appear to me in any way to " preclude the stale objection of " a want of variety, and a too frequent re-" currence of the same figures," which Sir Henry Steuart anticipates from their enlarged size : the figure is the same, whatever be its dimensions.

The sketch below represents the alteration I have ventured to make in " those elegant " forms, the oval, and the circle," which Sir Henry advocates.

The ideas here stated upon this subject being in direct opposition to those set forth in Steuart's Planter's Guide, and differing in some degree from the feeling manifested in the review of that work in the Quarterly, it becomes imperative to state the question at issue in such a point of view as will enable those who are interested in it to make their election. To this end it is necessary that the writers above mentioned should be heard at large upon a point, on which they have as unreservedly condemned the opinions just stated, as I have ventured to do those which they advocate.

The author of the Planter's Guide says,
" It is undeniably true, that there was great
" formality in the endless dotted clumps of
" Brown and his followers, which are long
" since exploded. Price alleged, with great
" severity and some truth, that a recipe could
" be given for making a place any where by
" Brown's system ; because you had only to
" take a belt with a walk in it, a few round
" clumps, and a formal piece of water, and
" the object was effected. But as to the cir-
" cular and oval clumps, as fashion always
" runs into extremes, it has now given us
" something greatly worse in their stead.

" It would have been nothing, after Brown
" (according to Price's witty remark) had
" changed *Quadrata Rotundis*, if the profes-
" sors of the present school had again sub-
" stituted *Rotunda Quadratis*, and restored the
" rectangular figures of a former day. But
" instead of this, our present landscape gar-
" deners have *made a merit*, and are regularly
" *vain* of disfiguring their most beautiful sub-
" jects with clumps and plantations, and
" even approaches, in the most zigzag and
" grotesque figures, which are ten times more
" hideous and unpicturesque than the worst

" productions of their predecessors ! As a
" late powerful writer says, ' Their plant-
" ' ations, instead of presenting the regular or
" ' rectilinear plan, exhibit nothing but a
" ' number of broken lines, interrupted circles,
" ' and salient angles, which are as much at
" ' variance with Euclid as with nature.   In
" ' cases of enormity, they have been made
" ' to assume the form of pincushions, of
" ' hatchets, of penny tarts, and of breeches
" ' displayed at old clothesmen's doors.' See
" Quarterly Review, No. 72.

" In all this they tell you they are *imitating*
" *nature!* they seem truly to be of opinion,
" that to *change* must be the same thing as
" to *improve;* and that, in order to display the
" taste of Price and Knight, they have only
" to reprobate that of Brown and Repton.
" There is no man, whose taste has been
" formed on any correct model, that does not
" feel and acknowledge the beauty of those
" elegant forms — the oval, the circle, and the
" cone — and who does not experience the
" pleasure of contemplating smooth and soft
" surfaces, every where marked by swelling
" undulations and gentle transitions.   Such
" are the outlines constantly prevalent in all

" the most beautiful objects in nature. We
" derive them originally from that most per-
" fect of all forms, the female figure; and
" there are few well educated persons who
" will for a moment compare to them a mul-
" titude of obtuse and acute angles, great
" and small, following each other in fantastical
" and unmeaning succession.

" If masses must be planted in parks, in
" order to get up wood for future single trees
" and detached groups, (which, without the
" interposition of the transplanting, they must
" be,) it is plain that they will continue in
" existence for five and twenty or five and
" thirty years, before they can be cut out with
" proper effect. What shape, I would ask,
" can be adopted with such distant objects
" in view, more generally pleasing than that
" of the circle, or the oval, or some modifica-
" tion of it? observing always, in laying out
" such plantations, to make the masses large
" enough, which will preclude the stale ob-
" jection of a want of variety, and a too fre-
" quent recurrence of the same figures. ' The
" ' man of taste' (as the eminent author above
" ' mentioned observes,) ' will be desirous that
" ' the boundaries of his plantations should

H

" ' follow the lines designed by nature, which
" ' are always easy and undulating, or bold,
" ' prominent, and elevated, but never stiff
" ' and formal.'

" It is to be hoped that there is discern-
" ment enough in our present race of artists
" to see the propriety of adopting or restoring
" those fine figures, the oval and the circle
" as certainly the best for temporary and
" large detached masses of wood.   And now
" that all controversy between hostile systems
" is at an end, I trust that the English garden,
" distinguished by simplicity and freedom,
" will henceforth be under no law but that of
" Nature, improved and embellished by such
" Art only as owns her supremacy, and knows
" to borrow, without being herself seen, every
" pleasing form which owes its origin to that
" unfailing source of variety and beauty." *

It appears singular that the advocates on
each side of the question before us, should
appeal to nature as the foundation of their
diametrically opposite systems.   I say advocates, as there are authorities for the view of
the subject which I have taken, at least as high
in matters of taste as either of those with

---

* Steuart's Planter's Guide, note 2. p. 422.

whom I have the misfortune to differ in this discussion; which authorities shall speak for themselves in due time.

The author of the Planter's Guide seems to me to have lost sight of nature altogether, as a model for our imitation in the subject before us, when he would lead us to " acknowledge " the beauty of those elegant forms, the oval, " the circle, and the cone." We do acknowledge them, and, with him, " experience the " pleasure of contemplating smooth and soft " surfaces, swelling undulations, and gentle " transitions;" and, with him, admire their beautiful prototype in the female form: we also most cordially agree in his following remark, that " there are few well educated " persons who will for a moment compare to " them a multitude of obtuse and acute angles, " great and small, following each other in " fantastical and unmeaning succession." We do, I repeat, most cordially agree with him in this position, as we cannot see what possible comparison can exist between them. Surely the smooth soft surface, the swelling undulations, and gentle transitions, exhibited by Nature in the most beautiful of all her works, the female figure, did not suggest the model for her rocks, her precipices, and her forests. As well might we compare a picture of Guido with one of Salvator Rosa, to adjust their separate excellence in their common art. Each took nature for his model: and the man of taste (who, I apprehend, is here meant by the well-educated,) will admire each without depreciating the other.

It will be remembered, that this system of circles and ovals is recommended for the planting of a park, or park-like scenery; and not only recommended, but insisted upon, as exclusively consonant with good taste. Let us hear what the same author says in another part of his book; where, in speaking of planting, he observes,—" But on such subjects, as " on most others connected with taste in the " disposition of wood, great diversity of opi- " nion must prevail; and that mode of ar- " rangement or execution will generally seem " the handsomest, in which the genius of the " place is best studied, and where the most " luxuriant growth and the most careless " disposition of wood are produced. The " greatest triumphs of Art must always be " those in which, in rivalling Nature, she most " completely effects her own concealment."

Could this just observation have been expected from the advocate for circles and
ovals? What agreement, let me ask him, can
exist between such monotonous forms and
the "*most careless dispositions of wood.*" How
is the *genius of the place* to be consulted in
the universal application of these *fine* forms?
or how is Art to effect her own concealment
under them? When, therefore, the author
of the Planter's Guide triumphantly asks,
" What shape can be adopted more generally
" pleasing than that of the circle, or the oval,
" or some modification of it?" he may be
answered, " Take any form but that." Neither is it at all apparent, that, however *large*
the masses may be, " the stale objection of a
" want of variety, and a too frequent recur-
" rence of the same figure, is any way re-
" moved;" as all the variety that can be given
will consist in the difference of size in these
monotonous forms; which forms being necessary, according to the writer's statement,
for twenty or thirty years, will never fully
escape from that thraldom: witness oval and
circular groups of full-grown trees in many
places worthy of better taste.

The paragraph which asks the above ques-

tion, "*What shape can be adopted so generally
"pleasing as the circle or the oval?*" concludes
thus: "The man of taste" (as the eminent
author above mentioned observes) " will be
" desirous that the boundaries of his plant-
" ations should follow the lines designed by
" nature, which are always *easy and undu-
" lating, or bold, prominent, and elevated, but
" never stiff and formal.*" How shall we reconcile this paragraph to itself? Did Nature
ever bound her plantations by a circular or
an oval form? Surely such forms are as
remote from the easy and undulating, as they
are from the bold and prominent character
of nature's outline; and must, I apprehend,
be classed under the " stiff and formal which
" she disowns."

I know not if the Planter's Guide intends
me the honour of a place among " our pre-
" sent landscape gardeners, who have *made a
" merit and are regularly vain* of disfiguring
" their most beautiful subjects with clumps
" and plantations, and even approaches in the
" most zigzag and grotesque figures, and
" which are ten times more hideous and un-
" picturesque than the worst productions of
" their predecessors :" the accusation appears

to admit of no exception; and is at least as severe as any thing the writer of it can find in Price, whose severity he censures. Where the author of the Planter's Guide has met with these *grotesque figures, these hideous and unpicturesque productions,* he has not told us; nor have I, in a tolerably extensive range of observation, discovered a single example of them.

The *conclusion* of this long note, as transcribed from the Planter's Guide, has, I confess, puzzled me extremely in my attempt to discover any support it affords to the object of the note itself,—the propriety of circles and ovals as applied to plantations. The passage runs thus :—

" It is to be hoped that there is discern- ,' ment enough in our present race of artists " to see the propriety of adopting or restor- " ing those fine figures, the oval and the " circle, as certainly the best for temporary " and large detached masses of wood. And " now that all controversy between hostile " systems is at an end, I trust that the En- " glish Garden, distinguished by simplicity " and freedom, will henceforth be under no " law but that of Nature, improved and em-

" bellished by such Art only as owns her
" supremacy, and knows to borrow, without
" being herself seen, every pleasing form
" which owes its origin to that unfailing
" source of variety and beauty."

Presuming that our present race of *artists*
means landscape *painters,* in contradistinction
to our present landscape *gardeners* so lately
denounced as destitute of all pretension to
taste, will the author of the Planter's Guide
forgive me if I say, it is to be hoped that
there is discernment enough in our present
race of artists to see the propriety of *omitting*
" *those fine figures the oval and the circle,*"
whenever they may be called upon to represent a scene disfigured by such misapplication of forms, though pronounced by him as
certainly the best for temporary and large
detached masses of wood?   I would ask, Are
these forms thus misapplied to be found in
the works of those artists, ancient or modern,
who have carried landscape painting to its
highest excellence?   Are they to be traced in
Claude or Poussin—in Wilson or in Turner?

Sir Uvedale Price entertains a much more
enlightened view of the question, when he
says,—" It may be said, with much truth, that

" the reformation of public taste in real land-
" scape more immediately belongs to the
" higher landscape painters, among whom
" the higher painters of every kind may
" generally be included; but there are cir-
" cumstances which are likely to prevent
" them from succeeding in a task for which
" they are so well qualified. In the first
" place, they have few opportunities of giving
" their opinion, being seldom employed in
" improved places; certainly not in repre-
" senting the improved parts: for there is a
" strong repugnance, of which the owners
" themselves are aware, in him who has studied
" Titian, Claude, and Poussin, and the style
" of art and of nature that they had studied,
" to copy the clumps, the naked canals, and
" no less naked buildings of Mr. Brown."*

It does not appear upon what grounds the
author of the Planter's Guide pronounces all
controversy between hostile systems to be at
an end, when he might himself be hailed as
the champion of the opposite opinions in the
subject before us. The Brownists would
triumphantly quote his recommendation of
those fine figures, the oval and the circle, as

* Price on the Picturesque, vol. ii. p. 179.

certainly the best models for their plantations;
while those who erect nature and the best
works of art for their standard, would transcribe upon their banner his concluding sen-
tence: — " I trust that the English Garden,
" distinguished by simplicity and freedom,
" will henceforth be under no law but that of
" Nature, improved and embellished by such
" Art only as owns her supremacy, and knows
" to borrow, without being herself seen, every
" pleasing form which owes its origin to that
" unfailing source of variety and beauty."

Before we enter upon the enquiry, to which
of the contending systems these observations
of the Planter's Guide lend their aid, I will
take the opportunity to disclaim, for myself at
least, all intention of controversy; for which
I have neither inclination nor leisure. When,
however, a work so widely circulated as is the
Planter's Guide—supported, too, by the powerful talent of its Reviewer,—when these authorities manifest such unreserved and sweeping
contempt of the principles which the study of
a long life has confirmed me in, I owe it to
my own professional character — I owe it to
those gentlemen, both in England and Scotland, who have honoured me with their ap-

probation — to show that I have not lightly undertaken the task they have severally committed to my care ; but that I have used my best endeavours to improve each place in conformity with its leading features, and to unite, as much as in me lies, every thing, from whatever source, that may tend to give propriety, character, and harmony to the whole. If I add, that the name I bear is not unknown as connected with subjects of taste, it is merely to suggest the probability that an early and long-continued intimacy with the relatives to whom I allude, would not leave me altogether uninformed of its true principles.

What then, we will ask, are those principles as applicable to our immediate subject — park or park-like scenery? The author of the Planter's Guide answers, — " Nature, im-
" proved and embellished by such Art only as
" owns her supremacy, and knows to borrow,
" without being herself seen, every pleasing
" form which owes its origin to that unfailing
" source of variety and beauty."

With this definition I cordially agree: it is the basis upon which I aim to found all my practice, — it is the model upon which I

venture to recommend the *irregular and
varied form* of planting, in preference to the
regular and monotonous oval and circle. Surely
Art, in administering to the embellishment of
a park, would seek in vain to *borrow* from the
scenery around her an authority for the
oval and the circle. If, therefore, she obtrudes
these forms, she no longer owns the *supremacy*
of *Nature;* but stands condemned, by the above
definition, as an handmaid devoid of all propriety and taste in pinning upon her mistress's
sylvan attire the ornaments that belong to
her robes of state and splendour.

It would appear that the predilecton for
ovals and circles arises from a too confined
view of the subject among its advocates. Their
great admiration of curved and flowing lines
prevents their investigation as to the propriety of their application; but if judicious
selection be a leading feature of good taste, it
is not easy to see how an application totally
foreign to the subject can have any foundation on that quality. Let the author of the
Planter's Guide spend a day amidst the
splendid scenery of the New Forest, or in
any of the natural woods I have visited in
Scotland, with a view to this question ; and I

am indeed mistaken, if the hand that has laboured so successfully for the embellishment of Nature in one particular line could have the hardihood—I had almost said the sacrilege—to insult her, thus enthroned, with a knot of circular or oval plantation. Let him, afterwards, behold in the late plantations in Richmond Park an example of the tasteless system he advocates — destitute of all variety — out of harmony with all around it.

A park, it will be allowed, is not a forest; but the " *genius loci* " is equally entitled to attention in our attempts at embellishing it: nor do I see how the " *easy and undulating* " *line of boundary*" can be produced by an assemblage of convex forms, whose only variety must arise from the difference of size and position.

Not to deprive the author of the Planter's Guide of any support to his system of circles, cones, and ovals, we will venture to examine the opinions manifested in the Review of his work ; for, though that masterly production does not absolutely prescribe the forms we are combating, yet I conceive it may appear tacitly to advocate them in the following passage : —

*" Repton, indeed, has justly urged, in fa-
" vour of the works of Kent and Brown, that
" the formal belts and clumps which they
" planted were intended only to encourage the
" rise of the young plantations, which were af-
" terwards to be thinned out into varied and
" picturesque forms; but which have, in many
" instances, been left in the same crowded
" condition and formal disposition which they
" exhibited at their being first planted.   If
" the school of Kent and Brown were liable
" to be thus baffled by the negligence of those
" to whom the joint execution of their plans
" were necessarily trusted, a much greater
" failure may be expected, during the subse-
" quent generation, from the neglect of plans
" which affect to be laid out upon the prin-
" ciples of Price.   We have already stated,
" that it is to be apprehended that a taste for
" the fantastic will supersede that which the
" last age have entertained in favour of the
" formal.   We have seen various efforts, by
" artists of different degrees of taste and
" eminence, to form plantations which are
" designed at some future day to represent
" the wild outline and picturesque glades of a

* Quarterly Review, March, 1828, p. 321.

" natural wood. When the line of these is
" dictated by the character of the ground,
" such attempts are extremely pleasing and
" tasteful. But when a bizarre and extrava-
" gant irregularity of outline is introduced
" upon a plane or rising ground; when its
" whole involutions resemble the irregular
" flourishes of Corporal Trim's harangue; and
" when we are told that this is designed to
" be one day a picturesque plantation; we
" are tempted to recollect the common tale
" of the German baron, who endeavoured to
" imitate the liveliness of Parisian society
" by jumping over stools, tables, and chairs
" in his own apartment; and when the other
" inhabitants of the hotel came to enquire the
" cause of this disturbance, answered them
" with the explanation *sh'apprends d'estre fif.*
" If the visiter applies to know the meaning of
" the angles and contortions introduced into
" the lines of the proposed plantations in
" Petruchio's language,

' What! up and down, carved like an apple tart;
Here's snip, and nip, and cut, and slish, and slash,
Like to a censor in a barber's shop;'

" he receives the plausible reply, that what

" he now sees is not the final result of the
" designer's art; but that all this fantastic zig-
" zaggery, which resembles the traces left by
" a dog scampering through snow, is but a set
" of preparations for introducing, at a future
" period, as the trees shall come to maturity,
" those groups and glades—that advancing
" and retiring of the woodland scene—which
" will realize the effects demanded by the
" lovers of the picturesque.   At present, we
" are told, that the scene resembles a lady's
" tresses in papillotes, as they are called, and
" in training for the conquests which they
" are to make when combed into becoming
" ringlets. But, alas! art is in this department
" peculiarly tedious; and life, as in all cases,
" precarious and short.   How many of these
" papillotes will never be removed at all,
" and remain unthinned out, like the clumps
" and belts of Brown's school, disfiguring the
" scenes they were designed to adorn."

I must here repeat, that examples in any
way agreeing with the above description have
not fallen under my observation: I will, however, meet them in all due deference to the
splendid talent of the author who states them
to exist; and, though I conceive that their

general bearing has been already discussed, I am anxious to examine the minutest point advanced by an authority so highly and so justly elevated in the annals of taste ; and I cannot but regret, that some extravagant imitations of nature's outline should have driven an eye so alive to the rich varieties of landscape scenery to adopt the dull monotony of the circular system in the accompaniments to that scenery.

It appears, then, I conceive, that the amount of the above critique is such as almost every thing in art or science is subject to ; viz. the mischievous effects of conceit and ignorance. But, surely, the value of medicine is not to be appreciated by the errors of empyricism, nor the art we are discussing to be estimated by the failures of those who are altogether igno-rant of the foundation on which it rests.

The study of nature, both in the disease and in the remedy, marks the skilful physi-cian ; the study of nature, in all her varieties of character and composition, can alone fit the man of taste for the supplying of her deficiencies, or the correction of her exuber-ances. The wise physician will improve his practical knowledge by the study of the best

authorities in medicine: the judicious improver will mature his taste by that of the
great masters in landscape painting. Empiricsin both these pursuits there undoubt-
edly will be, whose prescriptions will vary
from each other according to their own erroneous views of the subject. The practitioners
in landscape may be divided (to use the term
of the Review) between the fantastic and the
formal: but the affected and whimsical irregularity of the one is no commendation of
the dull monotony of the other; and, indeed,
more hope may be entertained of correcting
the extravagancies of the irregular system,
than of engrafting any improvement upon
that which has been pronounced as *certainly*
the *best*. At any rate, a happy result may
occasionally follow the labours of the most
irregular practitioner, while the circular remedy, though recommended in all cases, yet
being adapted to none, can produce no such
result, except as a warning to those who have
not yet suffered from its baleful effects.

But to return to the critique: I confess
myself somewhat puzzled in attempting to
substantiate the basis on which the author of
it rests his objection to the irregular form;

when, speaking of plantations so constituted, he observes,—" When the line of these is " dictated by the character of the ground, " such attempts are extremely pleasing and " tasteful. But, when a bizarre and extra- " vagant irregularity of outline is introduced " upon a plain or rising ground," &c. &c. Now, if neither a plain nor a rising ground be fit subjects for this irregularity of outline, I do not readily conceive that *character of ground, which renders such attempts extremely pleasing and tasteful.* At any rate, the blame is here divided between the form itself and its misapplication. I will venture, however, to suggest, that the irregular form is applicable to any character of ground, if intricacy and variety are essential to picturesque effect: and, indeed, a plain is, of all species of ground, perhaps, the most indebted to plantation for producing that effect; and whether, I will ask, are the qualities of variety, intricacy, and connection, to be sought in the irregular outline, or in that of the oval and circle? Upon this principle, I have treated the flat extent of park at Cassiobury, which, having been originally planted with similar groups of trees, afforded little of that variety of character, which, I trust, will result from the large irregular masses of plantation I ventured to recommend.

But to resume. The flourish of Corporal Trim's harangue will, perhaps, be conceded, as merely a flourish: but I have no hesitation in avowing my readiness to follow the out-line of the Reviewer's favourite, through his wildest vagaries over the snow, in preference to tracing the insipid formality of Kent or Brown; and I feel assured the result would justify the preference. The papillottes, from their similarity of size and shape, I conceive to be more allied to the circular system, than to the " zigzaggery," to which the Reviewer has attached them. The tale of the German baron is also, I think, at least as applicable to the dull uniformity of circles and ovals, claiming affinity with the playful elegance and variety of nature, as it is to the over-strained irregularity of the opposite system.

As a visible example is sometimes more convincing than any argument a slight illus-tration is here subjoined of the effect of the opposite systems, both in their infancy and their future maturity; and I should have no fear of the award of the authority with which

I am at present more immediately at variance, as I trust it will be obvious that the objections he has urged are not against the irregular outline, but against the conceit and ignorance he has seen manifested in its application.

The author of the Planter's Guide is not, I think, entitled to the support he assumes from the " late powerful writer" whom he quotes; and who, but for the term *late*, might be identified with the equally powerful writer from whom we have just parted. The passage on which he rests occurs in the Review of Monteith's Planter's Guide.* I have already given Sir H. Steuart's version of it : but, as the true bearing of the passage is not, I think, contained in that version, I beg leave to transcribe the whole passage itself, as most essential to the question before us.

The Review, having given some useful hints for providing the necessary plants, proceeds : " Thus provided with the material of " his enterprise, and with the human force " necessary to carry it into effect, the planter's " next point is to choose the scene of oper-

* See Quarterly Review, vol. xxxvi.

" ation. On this subject reason and common
" sense at once point out the necessary re-
" strictions. No man of common sense would
" select, for the purpose of planting, rich
" holmes, fertile meadows, or the ground
" peculiarly fit for producing corn or for
" supporting cattle.   Such land, valuable
" every where, is peculiarly so in a country
" where fertile spots are scarce, and where
" there is no lack of rough, exposed, and at
" present unprofitable tracts.   The necessary
" ornament of a mansion house would alone
" vindicate such an extraordinary proceeding.
" Nay, a considerate planter would hesitate
" to cut up and destroy even a fine sheep
" pasture for the purpose of raising a wood,
" while there remained on the estate ground
" which might be planted at a less sacrifice.
" The ground ought to be shared betwixt
" pasture and woodland, with reference to
" local circumstances ; and it is in general by
" no means difficult to form the plantation
" so as to be of the highest advantage to the
" sheep walk.   In making the selection, the
" proprietor will generally receive many a
" check on this subject from his land steward
" or bailiff, to whom any other agricultural

" operations are generally more desirable
" than the pursuits of the forester. To con-
" firm the proprietor in resisting this narrow-
" minded monitor, it is necessary to assure
" him, that the distinction to be drawn be-
" twixt the ground to be planted, and that
" which is to be reserved for sheep, is to be
" drawn with a bold and not a timid hand.
" The planter must not, as we have often
" seen vainly attempted, endeavour to ex-
" clude from his proposed plantations all
" but the very worst of the ground. When-
" ever such paltry saving has been attempted,
" the consequences have been very unde-
" sirable in all respects. In the first place,
" the expense of fencing is greatly in-
" creased; for in order to form these pinched
" and restricted plantations, a great many
" turnings and involutions, and independent
" fences, must be made, which become totally
" unnecessary when the woodland is formed
" on an ample and liberal scale. In the
" second place, this parsimonious system
" leads to circumstances contrary to Christian
" charity; for the eyes of every human being
" that looks on plantations so formed, feeling
" hurt as if a handful of sand were flung into

" them, the sufferers are too apt to vent their
" resentment in the worst of wishes against
" the devisers and perpetrators of such enor-
" mities. We have seen a brotherhood of
" beautiful hills, the summits of which, while
" they remained unplanted, must have formed
" a fine undulating line, now presenting
" themselves with each a circle of black fir
" like a skimming dish on its head, combined
" together with long narrow lines of the same
" complexion, like a chain of ancient forti-
" fications, consisting of round towers flank-
" ing a straight curtain, or rather like a range
" of college caps connected by a broad black
" riband. Other plantations, in the awk-
" ward angles which they have been made to
" assume, in order that they might not tres-
" pass upon some edible portion of grass
" land, have come to resemble Uncle Toby's
" bowling-green transported to a northern
" hill side. Here you shall see a solitary
" mountain with a great black patch stuck
" on its side, like a plaster of Burgundy
" pitch; and there another, where the plant-
" ation, instead of gracefully sweeping down
" to its feet, is broken short off in mid air,
" like a country wench's gown tucked

" through her pocket holes, in the days when
" such things as pockets were extant *in rerum*
" *naturâ*. In other cases of enormity, the
" unhappy plantations have been made to
" assume the form of pincushions, of hatchets,
" of penny tarts, and of breeches displayed
" at an old clothesman's door. These
" abortions have been the consequence of
" a resolution to occupy with trees only
" those parts of the hill where nothing else
" will grow; and which, therefore, is carved
" out for their accommodation with ' up and
" ' down and snip and slash,' whatever un-
" natural and fantastic forms may be thereby
" assigned to their boundaries.

" In all such cases, the insulated trees, de-
" prived of the shelter they experience when
" planted in masses, have grown thin and
" hungrily — affording the unhappy planter
" neither pleasure to his eye, credit to his
" judgment, nor profit to his purse. A more
" liberal projector would have adopted a
" very different plan. He would have con-
" sidered, that although trees, the noblest
" production of the vegetable realm, are of
" a nature extremely hardy, and can grow
" where not even a turnip could be raised,

" they are yet sensible of and grateful for
" the kindness which they receive.   In
" selecting the portions of waste land which
" he is about to plant, he would, therefore,
" extend his limits to what may be called
" the natural boundaries; carry them down
" to the glens on one side, sweep them around
" the foot of the hills on another, conduct
" them up the ravines on a third; giving them
" as much as possible the character of a na-
" tural wood, which can only be attained by
" keeping their boundaries out of sight, and
" suggesting to the imagination that idea
" of extent which always arises when the
" limits of a wood are not visible.   It is true,
" that in this manner some acres of good
" ground may be lost to the flocks, but the
" advantages to the woodland are a complete
" compensation.   It is, of course, in sheltered
" places that the wood begins to grow; and
" the young trees, arising freely on such more
" fertile spots on the verge of the plantation,
" extend protection to the general mass which
" occupies the poorer ground.   These less
" favoured plants linger long, while left to
" their own unassisted operations : annoyed at
" the same time by want of nourishment and

" the severity of the blast, they remain, in-
" deed, alive, but make little or no progress :
" but when they experience shelter from
" those which occupy a better soil, they seem
" to profit by their example, and speedily
" arise under their wings.

" The improver ought to be governed by
" the natural features of the ground, in choos-
" ing the shape of his plantations, as well as
" in selecting the species of ground to be
" planted. A surface of ground undulating
" into eminences and hollows, forms, to a
" person who delights in such a task, perhaps,
" the most agreeable of considerations on
" which the mind of the improver can be
" engaged. He must take care in this case
" to avoid the fatal error of adopting the
" boundaries of his plantations from the sur-
" veyor's plan of the estate, not from the
" ground itself. He must recollect that the
" former is a flat surface, conveying, after
" the draughtsman has done his best, but a
" very imperfect idea of the actual face of
" the country, and can, therefore, guide him
" but imperfectly in selecting the ground
" proper for his purpose.

" Having, therefore, made himself person-

" ally acquainted with the localities of the
" estate, he will find no difficulty in adopting
" a general principle for lining out his worst
" land. To plant the eminences, and thereby
" inclose the hollows for cultivation, is what
" all parties will agree upon: the mere farmer,
" because, in the general case, the rule will
" assign to cultivation the best ground, and
" to woodland that which is most sterile;
" and also, because a wood placed on an
" eminence affords, of course, a more com-
" plete protection to the neighbouring fields,
" than if it stood upon the same level with
" them. The forester will give his ready
" consent, because wood nowhere luxuriates
" so freely as on the slope of a hill. The
" man of taste will be equally desirous that
" the boundaries of his plantation should
" follow the lines designed by nature, which
" are always easy and undulating, or bold, pro-
" minent, and elevated, but never either stiff
" or formal. In this manner the future woods
" will advance and recede from the eye ac-
" cording to, and along with, the sweep of
" the hills and banks which support them,
" thus occupying precisely the place in the
" landscape where Nature's own hand would

" have planted them. The projector will
" rejoice the more in this allocation, that in
" many instances it will enable him to con-
" ceal the boundaries of his plantations; an
" object which, in point of taste, is almost
" always desirable. In short, the only per-
" sons who will suffer by the adoption of this
" system, will be the admirers of mathema-
" tical regularity, who deem it essential that
" the mattock and spade be under the pe-
" remptory dominion of the scale and com-
" pass; who demand that all inclosures shall
" be of the same shape and the same extent;
" who delight in straight lines and in sharp
" angles, and desire that their woods and
" fields be laid out with the same exact cor-
" respondence to each other as when they
" were first delineated upon paper. It is to
" be conjectured, that when the inefficiency
" of this principle and its effects are pointed
" out, few would wish to resort to it, unless
" it were an humorist like Uncle Toby, or
" a martinet like Lord Stair, who planted
" trees after the fashion of battalions formed
" into line and column, that they might assist
" them in their description of the battles of
" Wynendale and Dettingen. It may, how-

" ever, be a consolation to the admirers of
" strict uniformity and regularity, if any
" such there still be, to be assured that their
" object is, in fact, unattainable; it is as im-
" possible to draw straight lines of wood—
" that is, lines which shall produce the ap-
" pearance of mathematical regularity along
" the uneven surface of a varied country—as
" it would be to draw a correct diagram upon
" a crumpled sheet of paper, or lay a carpet
" down smoothly upon a floor littered with
" books.   The attempt to plant upon such a
" system will not, therefore, present the re-
" gular plan expected; but, on the contrary, a
" number of broken lines, interrupted circles,
" and salient angles, as much at variance
" with Euclid as with Nature."

Now, I will ask, is there any passage in
the whole of this quotation that warrants
Sir Henry Steuart's deduction from it? Are
the broken lines, interrupted circles, salient
angles, pincushions, hatchets, and penny tarts,
represented as the offspring of the *vanity* and
bad taste of our present landscape gardeners?
Is he borne out in his affirmation—" In all
" these they will tell you they are imitating
" nature? "   On the contrary, it appears, that

in his anxiety to claim the powerful support
of the writer whom he quotes, he has over-
looked that writer's own account of the origin
of the deformities which he has been cen-
suring, when he says,—" These abortions have
" been the consequence of a resolution to
" occupy with trees only those parts of the
" hill where nothing else will grow, and
" which, therefore, is carved out for their
" accommodation, with up and down, and
" snip and slash, whatever unnatural and
" fantastic form may be thereby assigned to
" their boundaries." We will, therefore, dis-
miss this whole group of uncouth forms, presuming only that the penny tarts were not of
the usual shape, or they would have better
suited the taste of the author of the Planter's
Guide.

So far, indeed, is the Review in question
from affording that assistance to Sir Henry
Steuart's system, which in truth it stands in
need of, that the spirit of the whole criticism
is as remote from Sir Henry's ideas, as those
ideas are from true taste. Let any one read
attentively the excellent hints as to the form
and disposal of plantations, which are to be
found both in the quotation just given, and

also throughout the whole of the Review; and it will be abundantly manifest, that the author of the Planter's Guide has mistaken not only the passages which he selects, but the general bearing of the whole.

Having already considered the only passage in the review of Sir Henry's own work* which in any way bears upon the question, I scruple not to say, that the whole spirit of that review, in strict conformity with the one on Monteith, appears altogether irreconcileable to the system of circles and ovals, as " *certainly the best for temporary and large* " *detached masses of wood.*" Nay, I am persuaded, that, were Sir Henry himself to visit some places originally planted in avenues and formal lines, he would be constrained to acknowledge the oval and circle to be as irreconcileable to such planting, as they are to the face and varied outline of Nature's hand.

Take, for instance, Burleigh; compare the circular and oval plantations of the outer park, with the original planting of that magnificent scenery. The original is formal; but that formality is accompanied by a grandeur

* Page 109.

that is in the strictest harmony with the mansion; and, though we cannot but occasionally regret the loss of some deep glade, by the intervention of a long line of trees, yet, owing to the number of these lines, and the variety of their situations, as well as of their length, many pleasing groups and happy combinations are produced as you pass through them. But where shall we find any grandeur, beauty, or variety, in the oval and circular clumps and plantations of the outer park, scattered here and there with no reference to each other, or the general character and scenery of the place —

" Marring fair Nature's lineaments divine ? "

Normington, in the same neighbourhood, exhibits a similar instance (though on a smaller scale) of the deformity of those fine figures, the oval and the circle, when applied to plantation.

ALTHOUGH the whole tenor of the authorities subjoined is in direct opposition to the circular system, a few of the passages which bear more immediately on the subject may be selected from them.

Whately, in his Observations on Modern Gardening, says, — " Though the surface of " a wood, when commanded, deserves all " these attentions, yet the outline more fre- " quently calls for our regard; it is also more " in our power ; it may sometimes be great, " and may always be beautiful. *The first re-* " *quisite* is irregularity. That a mixture of " trees and underwood should form a long " straight line, can never be natural ; and a " succession of easy sweeps and gentle rounds, " each a portion of a greater or less circle, " composing altogether a line literally ser- " pentine, is, if possible, worse. It is but a " number of regularities put together in a " disorderly manner, and equally distant from " the beautiful both of Art and of Nature. " The true beauty of an outline consists more " in breaks than in sweeps — rather in angles " than in rounds — in variety, not in succes- " sion. The eye, which hurries to the ex- " tremity of whatever is uniform, delights to " trace a varied line through all its intri- " cacies."*

Let us hear Sir Uvedale Price's opinion

* Observations on Modern Gardening, p. 42.

upon the circular system of planting: — " It
" must be remembered, that strongly marked,
" distinct, and regular curves, unbroken and
" undisguised, are hardly less unnatural and
" formal, though far less grand and simple,
" than straight lines; and that, independently
" of monotony, the continual and indiscrimi-
" nate use of such curves has an appearance
" of affectation, and of studied grace, which
" always creates disgust." * And again ;—" But
" the great distinguishing feature of modern
" improvement is the Clump— a name which,
" if the first letter were taken away, would
" most accurately describe its form and effect.
" Were it made the object of study how to
" invent something which, under the name
" of ornament, should deform whole districts,
" nothing could be contrived to answer that
" purpose like a clump. Natural groups
" being formed by trees of different ages and
" sizes, and at different distances from each
" other — often, too, by a mixture of those of
" the largest size with thorns, hollies, and
" others of inferior growth—are full of variety
" in their outlines: and from the same causes,

* Price on the Picturesque, vol. i. p. 231.

" no two groups are exactly alike. But
" clumps, from the trees being of the same
" age and growth, from their being planted
" nearly at the same distance, in a circular
" form, and from each tree being equally
" pressed by his neighbour, are as like each
" other as so many puddings turned out of
" one common mould. Natural groups are
" full of openings and hollows; of trees ad-
" vancing before or retiring behind each
" other; all productive of intricacy, of va-
" riety, of deep shadows, and brilliant lights.
" In walking about them, the form changes
" at each step : new combinations, new lights
" and shades, new inlets, present themselves
" in succession. But clumps, like compact
" bodies of soldiers, resist attacks from all
" quarters : examine them in every point of
" view — walk round them — no opening, no
" vacancy, no stragglers, but, in the true
" military character, *ils font face partout.*"
And the same author observes — " The mass
" of improvers seem, indeed, to forget that we
" are distinguished from other animals by
" being

' Nobler far, of look erect.'

" They go about

> ' With leaden eye that loves the ground;'

' and are so continually occupied with turns
" and sweeps, and manœuvring stakes, that
" they never gain an idea of the first prin-
" ciples of composition.

" Such a mechanical system of operations
" little deserves the name of an art.   There
" are, indeed, certain words in all languages
" that have a good and a bad sense; such as
" simplicity and simple, art and artful, which
" as often expresses our contempt as our ad-
" miration.   It seems to me, that whenever
" art, with regard to plan or disposition, is
" used in a good sense, it means to convey an
" idea of some degree of invention, of con-
" trivance that is not obvious; of something
" that raises expectation, and which differs
" with success from what we recollect having
" seen before.   With regard to improving,
" that alone I should call art, in a good sense,
" which was employed in collecting from the
" infinite varieties of *accident* (which is com-
" monly called Nature, in opposition to what
" is called Art), such circumstances as may
" be happily introduced, according to the

" real capabilities of the place to be im-
" proved. This is what painters have done
" in their art ; and hence it is, that many of
" these lucky accidents being strongly pointed
" out by them, are called Picturesque."*

Mason, in his Essay on Design in Gardening, in defining a clump, says, — " The word
" comprehends many regular (or nearly re-
" gular) figures of small plantations, whether
" square, (like Lord Shrewsbury's avenue of
" Clumps, in Oxfordshire,) circular, or oval,
" or approaching to either. The clumps
" alluded to in the text were chiefly regular,
" and mostly circular, and at that time ima-
" gined by me to have lost their vogue ; but
" I fear that they afterwards recovered it."

Knight, in contrasting the loose and varied
groups of nature with the formal clumps of
the improver, exclaims —

> " But, ah ! how different is the formal lump
> Which the improver plants, and calls a clump ! "

A writer, who was one of the first of those
who awakened the public attention to the
beauties of natural scenery, speaking of

* Price on the Picturesque, vol. i. p. 344.

plantations, says, — " Thus far we have con-
" sidered a *clump* as a *single independent*
" object — as the object of a *foreground,*
" consisting of such a confined number of
" trees as the eye can fairly include at once.
" And when trees strike our fancy, either in
" the wild scenes of nature, or in the im-
" provements of art, they will ever be found
" in combinations similar to these.

" When the *clump* grows *larger,* it becomes
" qualified only as a *remote object,* combining
" with vast woods, and forming a part of
" some extensive scene, either as a first, a
" second, or a third distance.

" The great use of the *larger clump* is to
" lighten the heaviness of a *continued distant*
" *wood,* and connect it gently with the plain,
" that the transition may not be too abrupt.
" All we wish to find in a clump of this kind
" is *proportion* and *general form.*

" With respect to *proportion,* the detached
" clump must not encroach too much on the
" dignity of the wood it aids, but must observe
" a proper subordination. A large tract of
" country covered with wood will admit seve-
" ral of these auxiliary clumps, of different
" dimensions. But if the wood be of smaller

" size, the clumps also must be smaller and
" fewer.

" With regard to the *general form* of the
" larger clump, we observed, that in a *single*
" tree we expected elegance in the parts.   In
" the *smaller clumps* this idea was relinquished,
" and in its room we expected a *general con-*
" *trast* in trunks, branches, and foliage.   But
" as the clump becomes larger, and recedes
" in the landscape, all these pleasingcontrasts
" are lost, and we are satisfied with a *general*
" *form*. No *regular* form is pleasing. A clump
" on the side of a hill, or in any situation
" where the eye can more easily investigate
" its shape, must be circumscribed by an irre-
" gular line, in which it is required that the
" undulations both at the base and summit of
" the clump should be strongly marked ; as
" the eye, probably, has a distinct view of
" both." *

Sir Uvedale Price, with his usual accurate
discrimination, says, — " It is only by a habit
" of observation, added to natural sensibility,
" that we learn to distinguish what is really
" beautiful from what is merely smooth and

* Gilpin's Forest Scenery, vol. i. p. 177.

" flowing, and to give a decided preference to
" the former." *

Upon the whole, it is, I think, obvious, that
the author of the Planter's Guide, having de-
voted much time and study to the maturing
of his ingenious work on the transplanting
of trees, has not exercised that *habit of observation by which we learn to distinguish what is
really beautiful from what is merely smooth and
flowing, so as to give a preference to the former.*
I conceive this is abundantly proved by the
discrepancy of his own observations upon the
subject, as above transcribed.

To the ovals and circles I would willingly
apply the observation with which the Review
dismisses a collection of forms scarcely more
at variance with taste. " We are happy to
" say that this artificial mode of planting,
" the purpose of which seems to be a sort
" of inscribing on every plantation that it
" was the work of man, not of nature, is now
" going fast out of fashion."

It cannot, however, be expected that the
author of the following observation can fully
appreciate the propriety and taste displayed

* Essays on the Picturesque, vol. i. p. 342.

in the passages selected from the Essays on
the Picturesque. " The Landscape, a poem
" by the ingenious Mr. Knight; and the
" Essays on the Picturesque, by that accom-
" plished scholar, Mr. Price, are productions
" of high merit, which we must ever value, as
" having been the means of retrieving the
" public taste, and showing what is unnatural,
" formal, or monotonous in the character of
" the school of Brown and Repton; yet, as
" these meritorious works were composed
" under peculiar circumstances, and during
" the bitterness of controversy, they should
" be read by the young student *cum grano*
" *salis.*   Mr. Loudon's able treatise on the
" ' Improvement of Country Residences,'
" (which came out in 1806, and has not been
" half so much praised as it deserves,) forms
" a far less exceptionable guide to the man
" of taste, or the country gentleman, who,
" having no practical skill himself, is yet
" desirous to improve real landscape where it
" exists, or to create it where it is wanting."*

Now, whatever be the merit of the treatise here recommended, to prefer it, or any
similar work, to the luminous and compre-

* Planter's Guide, p. 360.

hensive spirit that pervades the Essays on the Picturesque, as *a guide to the man of taste*, argues, surely, but a slender acquaintance with that quality. Neither are the *peculiar circumstances* under which the Essays were composed, and which impugn their authority, at all apparent; nor have I, at least, been able to discover the *bitterness of controversy* ascribed to them by Sir Henry Steuart. Mr. Repton's name, if I mistake not, occurs but twice in the whole work, and each time in a note. The first is upon the impropriety of breaking an avenue, where Sir Uvedale Price says, — " The bad consequence of this system " of separating trees which had long grown " together, is no where more apparent than " when an old avenue is broken into clumps; " yet it may very well happen that a land- " scape painter, however strongly he may " condemn the alteration, as it affected the " general views, and the character of the " place, might find some particular advantage " from it with respect to his own art : for, as " he is not obliged to make an exact portrait, " it is sufficient for his purpose if he discover " the principal materials for composition from " the spot where he places himself. He, there-

" fore, may select a view between any two of
" the clumps ; and as a very slight alteration
" in his expeditious art turns them into
" groups, the whole may form a very pleasing
" landscape : again, as only two of the clumps
" would appear, no one could suspect, from
" such a picture or drawing, that there were
" other clumps which strongly marked the
" old line of the avenue from every part
" where they were seen. All this is perfectly
" fair in the painter, with reference to his
" own art : but were he employed to show
" what would be the future effect of breaking
" an avenue into clumps, it would, in the
" same degree, be unfair; it would, in fact,
" be a deception, and tend to deceive his
" employer. Yet this is precisely what Mr.
" Repton has done, for the purpose of show-
" ing how an avenue may be broken with
" good effect. He has, in one plate, repre-
" sented the avenue on which the operation
" is to be performed, at its length, and of
" course describing the straight line; and,
" in common justice, he ought to have given
" the same view of it when broken : but he
" well knew what a figure his clumps would
" make when the straight line was dotted

" with them. He, therefore, in the other
" plate, has very dexterously changed both
" the point of view and the scale ; and as he
" knew that even a third clump would have
" marked the straight line, he has supposed
" himself at the exact point from which only
" two of them could be introduced into the
" drawing ; and to this painter-like liberty,
" he has added that of varying their forms, so
" as to give them some appearance of natural
" groups. Mr. Repton cannot be ignorant,
" that when trees have been long pressed on
" each side by others, whenever one or more
" of them are left separate, two of their sides
" must be naked and flattened ; and that
" although by degrees the nakedness is clothed
" with small boughs and with leaves, hardly
" any length of time will make the flatness
" completely disappear. This is what on such
" occasions ought fairly to be stated ; and, if a
" drawing or engraving be made, ought fairly
" to be represented : but it is singular, that
" the person who has most strongly written
" against the use of applying painting to land-
" scape gardening, should have furnished the
" most flagrant instance of its abuse."*

* Essays on the Picturesque, vol. i. p. 332.

The other instance, where Mr. Repton's name occurs, may be considered as complimentary rather than severe : —

" Mr. Repton, who is deservedly at the
" head of his profession, might effectually
" correct the errors of his predecessors, if to
" his taste and facility in drawing (an advan-
" tage they did not possess), to his quickness
" of observation, and to his experience in the
" practical part, he were to add an attentive
" study of what the higher artists have done
" both in their pictures and drawings. Their
" selections and arrangements would point
" out many beautiful compositions and effects
" in nature, which, without such a study,
" may escape the most experienced observer.

" The fatal rock on which all professed
" improvers are likely to split, is that of
" system : they become mannerists, both from
" getting fond of what they have done before,
" and from the ease of repeating what they
" have so often practised ; but to be reckoned
" a mannerist is at least as great a reproach
" to the improver as to the painter. Mr.
" Brown seems to have been perfectly satis-
" fied, when he had made a natural river look
" like an artificial one : I hope Mr. Repton

" will have a nobler ambition — that of hav-
" ing his pieces of water mistaken for lakes
" and rivers."*

Notwithstanding that Mr. Repton remained
unconvinced by the arguments of his powerful
opponent, he does not appear to accuse him
of any bitterness, or to manifest any towards
him, when he writes —

" SIR,

" I am much obliged by your attention in
" having directed your bookseller to send me
" a copy of your ingenious work. It has
" been my companion during a long journey,
" and has furnished me with entertainment
" similar to that which I have occasionally
" had the honour to experience from your
" animated conversation on the subject. In
" the general principles and theory of the art,
" which you have considered with so much
" attention, I flatter myself that we agree;
" and that our difference of opinion relates
" only to the propriety, or, perhaps, possi-
" bility, of reducing them to practice.

" I am obliged both to Mr. Knight and to
" yourself, for mentioning my name as an ex-

* Essays on the Picturesque, vol. i. p. 398.

" ception to the tasteless herd of Mr. Brown's
" followers. But while you are pleased to
" allow me some of the qualities necessary to
" my profession, you suppose me deficient in
" others; and, therefore, strongly recommend
" the study of what the higher artists have
" done both in their pictures and drawings ;
" a branch of knowledge which I have always
" considered to be not less essential to my
" profession than hydraulics or surveying,
" and without which I should never have
" presumed to arrogate to myself the title of
" Landscape Gardener, which you observe is
" a title of no small pretension," &c. &c.

The strongest passage in this letter is the
following : — " Amidst the severity of your
" satire against Mr. Brown and his followers,
" I cannot be ignorant that many pages are
" chiefly pointed at my opinions; although
" with more delicacy than your friend Mr.
" Knight has shown," &c.

The conclusion of the letter is in harmony
with the beginning :— " Notwithstanding the
" occasional asperity of your remarks on my
" opinions, and the unprovoked sally of Mr.
" Knight's wit, I esteem it a very pleasant
" circumstance of my life to have been per-

" sonally known to you both, and to have
" witnessed your good taste in many situ-
" ations. I shall beg leave, therefore, to
" subscribe myself, with regard and esteem,"
&c. &c.

I am the more solicitous to remove the
charge of *bitterness of controversy* from the
Essays, both as its supposed existence might
interfere with their utility, and also to rescue
the memory of their author from an imputation
which, from personal knowledge, I can testify
did no way attach to him. Whatever may appear severe throughout the work in question,
is attributable to an uncommon quickness of
perception, joined to a keenness of expression
that delighted all who had the pleasure of his
acquaintance, and who, I doubt not, would
universally confirm the appeal made to them
in the concluding part of his answer to Mr.
Repton's letter,—a letter which evinces that
a playful brilliancy of taste, not the *bitterness*
of controversy, guided the pen of an author,
who will be admired the more, the closer he
is studied.

The passage alluded to runs thus:—

" The joint compliment you have paid
" to my friend and me, I can, for my own

" part, return with great sincerity; and on
" this occasion I dare say I may answer for
" Mr. Knight. I fear, however, that as you
" complain of the occasional asperity of my
" supposed remarks on your opinions, you
" will not think me grown milder in this
" open and continued controversy; for, in
" the course of pointing out and explaining
" the tendency of many indirect attacks and
" insinuations, which at first sight might not
" be obvious, some degree of sharpness in my
" answer would naturally arise; but he who
" writes a formal challenge must not expect
" a *billet doux* in return. I may also observe,
" that every man (whatever the game may
" be) has his particular manner of playing;
" an allusion which may not unaptly be ap-
" plied to writing. I have been told by some
" of my friends, that my play is sharp; I
" believe it may be so; but were I to en-
" deavour to alter it, I could not play at all.
" I trust, however, that my friends will vouch
" for me, that whatever sharpness there may
" be in my style, there is no rancour in my
" heart.

" On reading over what I have written, I
" could not but lament that there should be

" any controversy between us. Controversy
" at best is but a rough game, and in some
" points not unlike the ancient tournaments ;
" where friends and acquaintance, merely for
" a trial of skill, and love of victory, with all
" civility and courteousness tilted at each
" other's breasts — tried to unhorse each
" other — grew more eager and animated
" — drew their swords — struck where the
" armour was weakest, and where the steel
" would bite to the quick, — and all without
" animosity. As these doughty combatants
" of the days of yore, after many a hard
" blow given and received, met together in
" perfect cordiality at the famous round
" tables ; so I hope we shall often meet at
" the tables of our common friends. And as
" they, forgetting the smart of their mutual
" wounds, gaily discoursed of the charms of
" beauty, of feats of arms, of various strata-
" gems of war, of the disposition of troops,
" the choice of ground, and ambuscades in
" woods and ravines—so we may talk of the
" many correspondent dispositions and stra-
" tagems in our milder art ; of its broken
" picturesque ravines, of the intricacies and
" concealments of woods and thickets, and

" of all its softer and more generally attrac-
" tive beauties.

" Though I have already, perhaps, dwelt
" too long on that great principle, connec-
" tion, yet I cannot conclude this letter with-
" out mentioning an example of its effects in
" a more important sphere. Not that its ef-
" fects are doubtful, but that it is an example
" by no means unapplicable to the subject on
" which I have been writing, and one that,
" in the present crisis, cannot be too much
" impressed on our minds.

" The mutual connection and dependance
" of all the different ranks and orders of men
" in this country: the innumerable but vo-
" luntary ties by which they are bound and
" united to each other (so different from
" what are experienced by the subjects of
" any other monarchy), are, perhaps, the
" firmest securities of its glory, its strength,
" and its happiness. Freedom, like the ge-
" neral atmosphere, is diffused through every
" part, and its steady and settled influence,
" like that of the atmosphere on a fine even-
" ing, gives at once a glowing warmth, and a
" union to all within its sphere: and although
" the separation of the different ranks and

" their gradations, like those of visible ob-
" jects, is known and ascertained, yet, from
" the beneficial mixture, and frequent inter-
" communication of high and low, that sepa-
" ration is happily disguised, and does not
" sensibly operate on the general mind. But
" should any of these important links be
" broken; should any sudden gap, any dis-
" tinct undisguised line of separation, be made,
" such as between the noble and the roturier,
" the whole strength of that firm chain (and
" firm may it stand) would at once be broken.

" May the strength of that exalted prin-
" ciple, whose effects I have so much enlarged
" upon, enable us to cultivate this and every
" other art of peace in full security, whatever
" storms threaten us from without; and as it
" so happily pervades the true spirit of our
" government and constitution, may it no
" less prevail in all our plans for embel-
" lishing the outward face of this noble
" kingdom,

' Till Albion smile
' One ample theatre of sylvan grace.'

" I will now conclude this long comment
" on your Letter; and as it is the first, so I

" hope it will be the last time of my address-
" ing you in this public manner : in every
" private intercourse and communication, I
" shall always feel great satisfaction.

" I am, &c. &c."

I trust that the passages above recited will
remove all imputation of *bitterness* from the
controversy between rivals, now alike indifferent to the meed of victory, alike uncon-
scious of the fair face of nature which
awakened the strife between them ; and that
the Essays, freed from every impediment to
their utility, will be considered (as they de-
serve to be) the standard of taste on the
subjects of which they treat ; and I shall
feel most gratified, if my humbler attempt
may prepare the uninitiated to reap the full
advantage of that elegant and interesting
work.

CHAP. V.

ON WATER. — THE ACCOMPANIMENTS OF IT.

Amongst all the beautiful objects of nature, there is none more interesting in itself, or more useful in the various combinations of landscape, than water: it possesses an universal attraction, and has ever been considered as the highest achievement of the improver's skill. In proportion, however, to the beauty of artificial water, when happily effected, is the difficulty of producing such a result. It will be obvious, that this difficulty will vary according to the natural character of the ground to be flooded, as the concealment of the head is the first thing to be attended to. When, therefore, the ground runs in an undulating valley, narrowing towards the end, the formation of the head becomes an easy process: but, where there is no such advantage of ground, the operation will require all the talent of the most experienced artist to construct his lakes or rivers so as to resemble those of nature; which resemblance can alone crown his labours with success.

In forming a piece of water, the first consideration will be the character it should assume; whether of river, pool, or lake. This consideration will be influenced by the size and character of the place, the shape of the ground, and the attendant circumstances of trees, &c. It may be well to remember that beauty, not quantity, is the object to be kept in view.

In water, as in a plantation, the outline is of the utmost moment; and the same observation will apply to both, viz. that the excellence of the form will depend upon the boldness of its indentations, not upon the frequency of their occurrence. These indentations should be formed with immediate reference to the house, if the water be seen from it; and care should be taken, that the remote bank or shore be not parallel with the house, as any depth of bay so situated will appear little better than a straight line, especially if the house does not occupy an elevated situation.

In staking out a piece of water, of whatever character, attention should be paid to the improvement of any variety of surface that may exist, as such hints will generally suggest better forms than can be made with the spade : where, however, no such varieties of surface are to be met with, they should be created by depositing the earth from the excavation, so as to give different degrees of elevation to the different points or promontories of the banks or shores of the river or the lake. A pool is a small lake.

The placing of these points or promontories will be decided from the principal stations from whence they are to be seen, so as not to be hereafter hidden by any mass of wood, or groups of trees, that may be judged essential to the general composition. It is desirable that these promontories should be marked with as bold a variety as the character and circumstances of the scene will warrant; and this boldness of contrast will be more easily effected, where the aid of trees (either existing, or to be planted) can be obtained, as it is easier to *hide* the junction of the created promontory with the ground beyond, than to *unite* them with good effect.* The planting,

* Mr. Repton seems to have mistaken the forming and the decoration of water, when he says, " treading of cattle " will soon give the banks all the irregularity they re- " quire." And again, " all rules for creating bushes

in such cases, should consist mainly of the lower growths; as they would more immediately, and more effectually, hide the junction of the artificial hillocks with the natural ground.

In the commencement of the operation which we have been considering, I should recommend (as in forming a plantation) the marking on paper the existing state of things, and adapting the situation, form, and circumstances of the water to be created, so as to produce *an harmonious whole;* infinitely more important than the size or even the beauty of the water, simply considered.

The general form of the water being staked out, the digging should not be taken too close to the form given, but at different places should be more or less within it, so as to give

---

" to enrich the banks are nugatory, except cattle are " excluded."

Better, I conceive, to be without water, than to have it thus stripped of all the circumstances upon which its interest depends. But where is the difficulty of protecting the partial planting of the banks requisite either to the river or the lake? There is a large piece of water now forming at Sudbrook Holm, near Lincoln, which, both in the irregularity of its banks, and in their decoration, will, I trust, prove Mr. Repton's theory altogether erroneous.

an opportunity for the water to form its own line against the bank. Parts also of the intended bank might be left as first broken down with the pickaxe, rather than be more determined by the spade. Upon the same principle, the earth should be so heaped upon the different hillocks as to allow room for it to fall irregularly towards the bottom, as nothing can be more unnatural than a hanging level, as the workmen term it.

Though the principal varieties of form will be obtained where the shallowness of the water admits best of the operation, yet, in forming the head, it is desirable to give some variety to its outline, instead of the straight line, or uniform curve, which usually characterize it. One bold promontory shooting into the water would divide the length of the head, especially where it is of considerable extent. Great improvement also might, I conceive, be made in the construction of the head, by giving variety to its surface, instead of making it a dead level, as is usually the case. This variety of surface would, moreover, give opportunity for planting near the water, which cannot be safely done on the level surface for fear of injuring the puddle bank.

It is seldom that the head can be constructed so as to unite easily with the ground beyond it; for which reason, the drive or walk should not (if it can be avoided) pass over it. Indeed, under no circumstances, should you be permitted to walk all round a piece of water, as, its limits being thus betrayed, its extent is ascertained; whereas, when the walk is so conducted as occasionally to come upon the water, and that at the best points of view, and to be constrained by the intervention of planting, &c., again to leave it, not only is the apparent extent, as well as the variety, greatly increased, but the wish to explore what is thus hidden creates an interest beyond any that complete disclosure could afford. The small but beautiful artificial lake at the Priory, near Stanmore, is an illustration of what has been here stated; where the form of the lake, the conducting of the walk, the beauty of the openings to the water, and the appropriateness and variety of the interposing masses, groups, and single trees, &c. afford a striking example of the correct taste that executed the whole, and which has also dictated the theory on which it was formed. This theory, as

detailed at large in the Essays on the Picturesque, will be read with great advantage
by those persons who wish to give to artificial
water the character of nature.

Had the piece of water in the pleasure
grounds of Buckingham Palace been formed
on such a model, the effect would have been
striking. The water, instead of being acces-
sible on all sides, would have been carried
under the woody base of the high mound in
the neighbourhood, and had been lost in a
deep recess of overhanging trees.*

The decoration of water will depend
much upon the general description of the
scenery around it. If that of the wild character of nature, the accompaniments of the
river or lake, should partake of the rough-
ness of that character. Broken banks, as
though indented by the action of the water, and
roots of trees bared by the same operation,
with their stem soccasionally slanting athwart
the stream, will unite the river, with corresponding boldness, to the scenery around. But
if the water reposes in the smoother lap of
nature, its decorations should be adapted to

* I speak of the water as it was when first made ; it may
have been improved.

the tranquillity of the scene. Decoration, however, it must have, as an uniform bank or shore will never assimilate the artificial river or lake to their prototypes in nature's works. The banks of the river, though not so boldly contrasted as in the wilder scene, will still admit of considerable variety in their decoration. The smooth grassy bank, sliding almost imperceptibly into the water, will be relieved by a jutting point fringed with the varieties of water plants, enriched, if circumstances allow, with fragments of stone of different size and colour: groups of alder or willow will occasionally break the margin, or the pendent and massive foliage of the wych elm will throw its broad shadow across the retiring reach. These and such like circumstances are essential to the completion even of the happiest form that the artificial river can assume.

Mr. Repton appears to abandon all attempt at the imitation of nature's lakes or rivers, when he says, " Mr. Price has written an " Essay to describe the *practical* manner of " finishing the banks of artificial water: but " I confess, after reading it with much atten- " tion, I despair of making any practitioner

" comprehend his meaning; indeed, he con-
" fesses that no workman can be trusted to
" execute his plans. It is very true, that large
" pieces of water may be made too trim and
" neat about the edges; and that often, in
" Mr. Brown's works, the plantations are not
" brought near enough to the water; but if
" the banks are finished smoothly at first, the
" treading of cattle will soon give them all the
" irregularity they require: and with respect
" to plantations, we must always recollect,
" that no young trees can be planted without
" fences, and every fence near the water is
" doubled by reflection; consequently, all
" rules for creating bushes to enrich the
" banks are nugatory, except where cattle
" are excluded."*

Now, what are the difficulties that so alarm
Mr. Repton, as to deter him from the imitation of that which, in the same page, he allows
to be beautiful? " There is a part of the river
" Teme, above the house, where both its
" banks are richly clothed with alders; and
" every person of discernment must admire
" the beauty of this scene."

* Observations on the Theory and Practice of Landscape Gardening, p. 119.

However difficult it may be to imbue the mind of the workman with the principles of taste, the professor, possessing those principles, is surely able to direct the hand of the labourer in every practicable improvement.

Whether in creating the varied forms of the banks themselves, or in clothing them when formed, the operation, as far as the workman is concerned, differs nothing from labours in any other part of improvement: and I conceive it to be the fault of the professor, if the "*smoothly finished banks*" are left to the tread of cattle, to " give them all the " irregularity they require."

The doubling of the fence by reflection in the water deters Mr. Repton from planting the banks. As reflection is one of the greatest charms of water, this apprehension is ill-timed, when it would expose the broad ample channel to the sky, and leave a naked margin round it.

In like manner, the artificial lake must be united with the surrounding scenery; and the wood, which adorns its margin, must be connected, by means of groups of trees of various size and character, with other masses in its neighbourhood, either already existing, or to be planted for this purpose of connection. A sandy or gravelly promontory shooting into the water is very useful in producing a variety of colour, and giving a focus of light. A boat-house, or fishing-cottage, when aptly placed, are, upon the same principle, favourable circumstances on the border of the lake, which, like the bank of the river, will be enriched with the different plants suited to it. These plants, however, should not be universally introduced along the margin; as a line of uninterrupted material, of whatever kind, would destroy that variety, which is the very essence of the beauty of a lake.

A lake is more easy of successful imitation than a river, as greater variety of outline can be given, upon a moderate extent, to the former, than to the latter. The lake, also, may be complete, though on a small scale; whereas the artificial river can only be, comparatively, a small portion of the character it assumes. The difficulty of concealing the extremities of the artificial river, so as to impress the idea of continuity, will be considerable, even under the most favourable circumstances; add to which, models for lakes pre-

M

sent themselves upon a scale that admits of entire imitation. Ponds on commons, where the ground is of unequal surface, frequently assume the varied shores of a beautiful lake. Settlements of water in old pits suggest admirable hints to the same purpose; wanting only an increased extent, and judicious plant ing, to form a complete lake.

Islands, if properly introduced, and well formed, are essentially useful in creating variety of composition, as the lake is viewed from different situations. The size and number of them must depend upon the size of the water, and the circumstances of the place. It is hardly possible that an island in the middle of a lake can have a good effect; neither can a regular form convey the idea of a natural island. The effect of an island will sometimes depend upon its being raised above the level of the neighbouring shore, by which means greater variety and intricacy will be produced. It is also frequently desirable, that the island itself should be of unequal height. A low point projecting from under the hanging wood of the higher part of the island, especially if the point can be enriched with fragments of stone, or varied in its colouring by the warm tints of a sandy or gravelly soil, will have the happiest effect.

In the description of a scene in nature by two persons equally alive to its beauties, the general features of the composition would appear pretty much the same in each; yet, in the detail, considerable variety might be found; the attention of one having, perhaps, been principally directed to the beauty and harmony of the colouring and general effect, while that of the other had been more immediately occupied in tracing the intricacy, variety, and elegance of the outline. But, if these two persons were required to explain the principles upon which such a scene might be successfully imitated, being then under the necessity of critically examining all the component parts which constituted its beauty, the only difference between them would arise from the superior taste and discernment that either might possess above the other: their rule of operation would be the same in practice.

In such situation am I just now placed; agreeing entirely with Sir Uvedale Price through the whole of his discussion on the Picturesque; admiring and studying equally

with him the works of Nature and of Art,
with a view to the establishing a just basis for
the improvement of Landscape Gardening; I
have hitherto, I trust, treated the subject in
unison with him, without following his track.
Sir Uvedale has given to the public the fruit
of his research, adorned with that brilliancy
of talent, which so eminently distinguished
him. I have endeavoured to suggest the
same principles in humbler attire; but, having arrived at the most difficult point of
landscape improvement, which consequently
requires the most minute attention to the
detail, viz the decoration of artificial water, I
find Sir Uvedale's collection of circumstances
so comprehensive, and their application so
complete, as to leave me no choice but of
appearing to repeat his sentiments in my
words, or to let mine be given in his. Secure
of the approbation of the reader, I shall adopt
the latter, and transcribe part of the " Essay
on Artificial Water," &c.

" Islands in artificial water have, in many
" instances, been so shaped, and so placed, as
" to throw a ridicule on the use of them;
" but, if we once allow ourselves to argue
" from abuse, they would not be the only

" imitations of natural objects that ought to
" be condemned.   That islands are often
" beautiful in natural scenery, and in a high
" degree productive of variety and intricacy,
" can not be doubted ; and if it be true, that
" those parts of seas and large lakes where
" there are most islands (such as the entrance
" of Lake Superior or the Archipelago) are
" most admired for their beauty, and if the
" manner in which those islands produce
" that beauty be by dividing, concealing, and
" diversifying what is too open and uniform,
" the same cause must produce the same
" effect in all water, however the scale may
" be diminished; the same in a pool or a
" gravel pit as in an ocean.

  " Islands, though very common in many
" rivers, yet seem (if I may be allowed to say
" so) more perfectly suited to the character
" of lakes ; and, as far as there is any truth
" in this idea, it is in favour of making the
" latter our chief models for imitation.   In
" artificial water, the most difficult parts are
" the two extremities, and particularly that
" where the dam is placed; which, from
" being a mere ridge between two levels, is
" less capable of being varied to any degree

M 3

" by bays and projections, or by difference
" of height. The head, therefore, must, in
" general, be the most formal and uninterest-
" ing part, and that to which a break, or a
" disguise of some kind, is most necessary;
" but as it is likewise the place where the
" water is commonly the deepest, neither a
" projection from the land, nor an island, can
" easily be made thereabouts.* There are
" generally, however, some shallow parts at
" a sufficient distance from the head, where
" one or more islands might easily be formed,
" so as to conceal no inconsiderable portion
" of the line of the head from many points.
" In such places, and for such purposes,
" islands are peculiarly proper; a large pro-
" jection from the side of the real bank,
" might too much break the general line;
" but by this method that line would be pre-
" served, and the proposed effect be equally
" produced.

" It is not necessary that islands should
" strictly correspond with the shores, either in
" height or shape; for there are frequent
" instances in nature, where islands rise high

---

* I do not see this difficulty in so strong a light, and
have given my ideas upon it in page 155.

" and abruptly from the water, though the
" shore be low and sloping; and this liberty
" of giving height to islands may be made
" use of with particular propriety and effect
" towards the head, which usually presents a
" flat thin line, but little disguised or varied
" by the usual style of planting. An island,
" therefore, (or islands, as the case may re-
" quire,) in such a situation as I have pro-
" posed, with banks higher than those of the
" head, abrupt in parts, with trees projecting
" sideways over the water, by boldly advanc-
" ing itself to the eye, by throwing back the
" line of the head, and showing only part of
" it, would form an apparent termination of a
" perfectly new character; and so disguise
" the real one, that no one could tell, when
" viewing it from the many points whence
" such island would have its effect, which was
" the head, or where the water was likely to
" end.

" In forming and planting these islands, I
" should proceed much in the same manner
" as in forming the outline of the other
" banks. I should stake out the general
" shape, not keeping to any regular figure,
" and then direct the labourers to heap up

M 4

" the earth as high as I meant it should be,
" without levelling or shaping it, making
" allowance for its sinking, and reserving
" always the best mould for the top.   In the
" course of heaping up the earth without
" sloping it, a great deal will fall beyond the
" stakes, and would unavoidably give some-
" thing of that irregularity and play of out-
" line which we observe in natural islands ;
" the new earth would likewise settle, and
" fall down in different degrees, and in vari-
" ous places ; from all which accidents in-
" dications how to give greater variety might
" be taken.   If it be allowed that a mixture
" of the lower growths is as generally useful
" as I have supposed, it must be particularly
" so in islands, where partial concealment is
" so principal an object ; and as you can
" never give such a natural appearance of
" underwood and of intricacy, can never so
" humour the ground, so mark its varieties,
" especially on a small scale, by planting as
" by sowing,—it is most advisable to plant
" only what is more immediately necessary,
" and to sow seeds and berries of the lower
" growths, quite from the lowest growths of
" all ; and to encourage fern, and whatever

" may give richness and naturalness. In any
" part where I wished the boughs to project
" considerably over the water, I would raise
" the bank higher than the rest of the ground,
" and many times give it the appearance of
" abruptness; yet, by means of stones and
" roots, endeavour both to render it pictu-
" resque, in its actual state, and to prevent any
" change from its being broken down. On
" this high point I should plant one or more
" of such trees as had already an inclination
" to lean forward, from having been forced
" in that direction by trees behind them;
" and some of that kind are generally to be
" met with even in nurseries and plantations.
" By this method the bank and the trees of
" that part of the island would have a bold
" effect; and in places where the water began
" to deepen so much that it would be dif-
" ficult to extend the island itself any farther,
" its apparent breadth, and consequently the
" concealment occasioned by it, would in no
" slight degree be extended.

" The best trees for such a situation are
" those which are disposed to extend their
" lateral shoots, and are not subject to lose
" them by decay, and which likewise will

" bear the drip of other trees; such, for in-
" stance, as the beech, hornbeam, witch elm,
" &c.; or should the insular situation, not-
" withstanding the height of the bank, be
" found too moist for such trees, the improver
" will naturally choose from the various
" aquatics what will best suit his purpose.
" Among them the alder, however common,
" holds a distinguished place, on account of
" the depth and freshness of its green, and
" its resemblance, when old, to the noblest
" of forest trees—the oak. In a very dif-
" ferent style the plane is a tree of the most
" generally acknowledged beauty; and it may
" be observed, that the boughs, both of that
" and of the witch elm, form themselves into
" canopies, with deep and distinct coves be-
" neath them, in a greater degree than those
" of almost any other deciduous trees; a
" form of bough peculiarly beautiful when
" hanging over water. As the aim of the
" planter would be to make the whole of
" these trees push forward in a lateral direc-
" tion, it might often be right to plant some
" other trees behind them of a more aspiring
" kind, such as the poplar; and by means of
" such a mixture, together with some of the

" lower growths, very beautiful groups may be
" formed, without any appearance of affected
" contrast.

" It may not be useless to remark on this
" occasion, that all trees, of which the foliage
" is of a marked character, and the colour
" either light and brilliant, or in the opposite
" extreme, should be used with caution, as
" they will produce light or dark spots, un-
" less properly blended with other shades of
" green, and balanced by them. The fir tribe,
" in general, has not a natural look upon
" islands on a small scale; but should a mix-
" ture of them happen to prevail on the
" other banks of the water, the cedar of Li-
" banus would remarkably suit the situation
" I have just mentioned; and that, and the
" pine-aster, in place of the poplar rising
" behind it from amidst laurels, arbutus, &c.,
" would form altogether a combination of
" the richest kind.

" All the plants which I have hitherto
" mentioned, are such as take root on dry
" land, or at least above the surface of the
" water; but there are others which grow
" either in the water itself, or in ground ex-
" tremely saturated with moisture, and there-

" fore must, of course, be suited to the char-
" acter of islands. These are the various sorts
" of flags, the bull-rush, the waterdock, &c.;
" to which may be added, those plants which
" float upon the surface of the water, such as
" the water-lily. From the peculiarity of
" their situation and of their forms, and from
" the richness of their masses, they very much
" contribute to the effect of water, and great
" use may be made of them by a judicious
" improver, particularly where the shore is
" low. I have observed a very happy effect
" from them in such low situations towards
" the extremity of a pool, that of preventing
" any guess or suspicion where the water was
" to end, although the end was very near.
" This is an effect which can only be pro-
" duced by islands, or by such plants as root
" in the water; for where trees or bushes
" grow on low ground, however completely
" they may conceal that ground by hanging
" over the water, yet we know that the land
" must be there, and that the water must end;
" but flags or bull-rushes being disposed in
" tufts or groups behind each other, do not
" destroy the idea of its continuation.
    " A large uniform extent of water, which

" presents itself to the eye, without any in-
" tricacy in its accompaniments, requires to
" be broken and diversified like a similar
" extent of lawn ; though by no means in the
" same degree : for the delight which we
" receive from the element itself, compen-
" sates a great deal of monotony. Islands,
" when varied in their shape and accom-
" paniments, have the same effect as forest
" thickets, circular islands that of clumps ;
" and the same system which gives rise to
" round distinct clumps, of course, produces
" islands equally round and unconnected. As
" the prevailing idea has been to show a
" great uninterrupted extent, whether of grass
" or of water, islands on that account have
" been but little in fashion : I have, indeed,
" seldom seen more than one in any piece of
" artificial water, and that apparently made
" rather for the sake of water-fowl than of
" ornament. When one of these circular
" islands is too near the shore, the canal
" which separates them is mean, and the
" island from most points appears like a pro-
" jection from the shore itself ; and when, on
" the other hand, it is nearly in the centre
" (a position of which I have seen some very

" ridiculous instances), it has much the same
" unnatural, unmeaning look, as the eye
" which painters have placed in the middle
" of the Cyclops' forehead; and that is one
" of the few points on which the judgment
" of painters seems to me to be nearly on a
" level with that of gardeners: they have
" an excuse, however, which I believe the
" latter could never allege—that of having
" been misled by the poets."

In the above quotation I have selected
what more immediately applies to the decoration of the banks: the whole of the Essay
I have already recommended to the study of
every one, who is desirous of forming a piece
of artificial water.

Before quitting this part of our subject, it
may not be amiss to suggest great care in in-
terfering with the character of a brook. Where
the ground and other circumstances concur,
the stream may occasionally be brought to
spread itself into a little pool; its indefinite
margin of alder, willow, and other bushes on
the lower side concealing the resumption of
its modest channel, till some favourable opportunity may again allure it from its retirement; thus creating a variety without

destroying its character. But it should be well considered, before the brook is sacrificed for a piece of water, whether the latter can be so formed and decorated as to warrant the change.

CHAP. VI.

MISCELLANEOUS.

ENTRANCE TO THE HOUSE NOT TO BE ON THE SIDE OF
THE LIVING ROOMS. — ALTERING THE ENTRANCE. —
TURNING THE APPROACH. — REMOVING HEDGE ROWS. —
CAUTION NECESSARY. — KITCHEN GARDEN. — ON OPEN-
ING SCENERY. — CHEERFULNESS AS CONNECTED WITH
SCENERY. — BLUE DISTANCE ONE GREAT SOURCE OF IT.
— A VILLAGE OR HAMLET CONDUCIVE TO IT. — MIS-
TAKE IN HIDING SUCH CIRCUMSTANCES. — PLANTING.
— CHARACTERS OF TREES. — SCOTCH FIR. — VARIOUS
OPINIONS OF IT. — TREES IN GROUPS. — LEVELLING AND
SMOOTHING. — HYDE PARK. — FENCES. — LODGES. —
BRIDGES. — VILLAGES.

IN the arrangement of the foregoing chapters,
I have endeavoured, as far as I have been
able, to keep the connection free from interruption; to effect which I have reserved
several insulated questions, as they may be
termed, for a miscellaneous discussion: and,
though the bearing of some of them may,
perhaps, have been already partially noticed,
yet, having, in the Introduction, requested
the reader's indulgence for such repetition,
and utility being the end proposed, I will

proceed to consider such questions as may occur to me.

It has been already stated, that judicious improvement must be founded upon the size, character, and circumstances of the place to which it is applied. Were this rule more generally observed, the result would be a harmonious consistency in each place, and a variety when compared with others of even apparently similar features. There is one circumstance, however, which, in my opinion, is equally applicable to all,—from the palace to the smallest residence of gentility,—viz. the entrance. I have, in another place, strongly expressed my feeling upon this subject; but, as it is an error, not of accident, but of design, I cannot but press the consideration of it as a matter of the utmost moment, both to internal comfort, and to external effect.

Where a house is to be built, I would request the owner to study well the scenery around; for want of which precaution we frequently find the offices occupying the ground on which the drawing-room should have been placed, and the entrance destroying the repose of the library. Where a house

N

is already so unfortunately circumstanced, I
should recommend the trying of every possible method to remedy the evil.   I have suc-
ceeded in several instances with houses of various size and character; and I think there are
comparatively but few that might not be thus
improved.   Examples of this improvement
will be found at Hawarden Castle, near Ches-
ter, Castleton, near Carlisle, and at Wickham
Park, in the vicinity of Croydon.   And even
where circumstances may not admit of the
entrance being changed, the approach may
frequently be so conducted as to become
less exposed to the living-rooms: sometimes
by taking it round the back of the house,
instead of crossing the front of it; sometimes
by merely altering the back road a considerable improvement may be effected.   The
wrong situation of the entrance is the only
blot in the beautiful scenery at Marston, in
Somersetshire; but I am not without hope
of seeing it corrected.   A great improvement
has been effected at Mell's Park in the same
neigbourhood, by removing the road to the
stables, which passed the library windows,
and contracted the shrubbery within a limit
incompatible with the size of the mansion.

Part of that road is now converted into a
handsome terrace, commanding, through the
trees of the lofty bank which it surmounts, a
fall of water backed by a rising wood on the
other side of the valley through which the
river glides. The whole of this interesting
scenery was completly excluded from the
dress-ground by the injudicious situation of
the stable-road. Speaking of the entrance,
I would observe, that, in my opinion, there is
scarcely any circumstance which can justify
the driving round a plot of grass, either naked
or planted with shrubs: neither would I generally make a semicircular sweep of gravel
before the door; a rectangular form being
more in harmony with the architectural arrangement of the building, especially if it be
on a large scale.

As a mansion, of whatever character, requires a corresponding extent of domain, it
will, in forming this accompaniment, be frequently necessary to remove the limitation
of hedge-rows, to change arable into pasture, and to clothe the widened extent with
large masses of plantation. But the dwelling
of less pretension by no means requires this
sacrifice. Here the approach may cross a

field or two, through an irregular kind of
avenue, to avoid gates; the arable may retain
its situation; the hedge-rows may be kept
up without detriment to the character of the
house, and its immediate accompaniments.
So also, I conceive, that the stable offices,
and even the farm buildings, if well grouped
with trees, and not in the way of the view,
may frequently be retained with perfect propriety in connection with a house of this
description, of which shelter is an indispensable characteristic. One or two openings to
the country, made with judgment, will generally be preferable to a more extensive clearing, especially on the approach side.

It may be here suggested, that, where extent of park or sheep-walk is necessary, great
caution should be exercised in the removing
of hedge-rows. How often do we see a line
of trees standing each on a little mound, and
marking the course of the hedge in which
they had stood; whereas, had the hedge been
partially removed, including the trees in that
removal, and in places the hedge and trees
been left standing together, a group of thorns,
or sometimes a single tree, planted so as to
break the straight line, would obliterate all

vestige of a former separation. When, from the beauty of the tree itself, or from any other cause, one should be retained in the opening, the mound should be softened off as gently as possible, and a thorn either left or planted to break the swelling line on one side. A good effect may frequently be obtained by planting the angles of cross hedges, whether the hedges are to remain or not. In the first case, such planting gives a general appearance of wood to the scene at small expense of fencing; in the second instance, good groups of trees may be obtained.

The placing of the kitchen garden is frequently a question of great difficulty; but in this, as in the case of the offices, much will depend upon the size and character of the house. Where an easy connection between the dress-ground and the kitchen-garden can be obtained without detriment to the scenery, I should recommend such an arrangement; as a kitchen garden so situated offers an agreeable variety of interesting circumstances, and may furnish flowers for the decoration of the house, without robbing the beds in the dress-ground. This situation of the kitchen-garden may occasionally be compatible with the character and extent of a mansion,—particularly of the manorial character,—but might frequently be adopted with great propriety in connection with the house of less pretension. The kitchen-garden offers an opportunity for a straight broad walk, should circumstances not admit it in the dress-ground.

There is no part of improvement in which cautious operation is so necessary as in opening scenery. A desire of extent and a love of prospect have done irreparable mischief in numberless instances. Injudicious *planting* may be remedied; but the evil resulting from injudicious *removal* can never be repaired, at least in the lifetime of the owner who has fallen into this common and destructive error. I was once consulted upon the improvement of a place on a large scale. The striking fault was, the want of trees on the foreground connected with the house: I therefore marked several places for groups and single trees to supply this deficiency. But what was my astonishment, when the owner told me he had cut down the trees that had occupied the very situations I had selected for planting.

It is under such circumstances, that a knowledge of landscape-painting is peculiarly applicable. Such knowledge would teach the necessity of studying the character of the surrounding country, and the impropriety of destroying the rich embowered scenery of Hobima, in the vain hope of obtaining the graduated and aërial distance of Claude.

Cheerfulness, as connected with scenery, being generally the object proposed in this indiscriminate clearing, it may not be un-instructive to examine into the component parts, if we may so speak, of this quality. It has been already stated, that blue distance, from its susceptibility of change under the variation of sun and cloud, and of the different periods of the day, &c., offers a perpetual subject of investigation to the eye, and hence constitutes one main source of cheerfulness in scenery.*

* Mr. Repton appears to be insensible to this cause of cheerfulness when he says, " But as distant prospects de-" pend so much on the state of the atmosphere, I have " frequently asserted, that the views from a house — and " particularly those from the drawing-room—ought rather " to consist of objects which evidently belong to the place." And again, " Views of distant mountains, which may be " seen as well from the high road, are not features that " justify extensive lawn over a flat surface."

As, however, all places have not the advantage of blue * distance, we must seek
for other causes to enliven such scenes. A
hamlet or village partially seen through the
accompanying trees, presenting a variety of
form and colour to the eye, and suggesting
many a pleasing reflection to the mind, will
imperceptibly spread a cheerful hue on all
around. Even the curling smoke rising from
the lonely cottage, and slowly floating across
the darkening wood below it, marking the pre-
paration for the labourer's evening meal, cannot but awake a kindly social feeling, and
impart a conscious cheerfulness to the mind
of the beholder.

I lately met with a most striking instance
of excluding such rural circumstances. A
mansion of the manorial character, commanding a rocky gorge, fringed with wood, through
which a river forces its agitated course, presents
to the library window a truly romantic scene,
of which a group of trees, on a precipitous
bank, about fifty feet from the house, forms
the foreground.

At the mouth of the gorge stands a picturesque cottage, as if placed by the hand

* This term is used to signify that distance which melts
into the horizon.

of taste itself. The improver, conceiving this cottage an improper appendage, has removed a large portion of earth (from a situation to which it was essential) to raise a mound on this foreground, for the purpose of excluding it : this he has completely effected, and, with the cottage, has shut out the valley, the gorge, the river, and buried behind his mound the boles of the foreground trees; thus contriving to render abortive the judicious selection of the architect. A more glaring example of perverted taste cannot exist.

Shenstone observes, " A rural scene, to me, " is never perfect without the addition of " some building. A cottage is a pleasing ob- " ject, partly on account of the variety it " may introduce, on account of the tran- " quillity that seems to reign there, and, " perhaps, I am somewhat afraid on account " of the pride of human nature."

The improver above mentioned seems to have been alike insensible to both the causes of Shenstone's predilection for the cottage in landscape-scenery.

But, perhaps, life and motion are, after all, the principal sources of cheerfulness, as connected with scenery. The sea is always *grand;* but it is the varying circumstances of navigation which imparts *cheerfulness* to the scene. This will be obvious to every one who has observed the contrast between Muddiford and Dover; as also in many other places on the sea-side.

This necessity of life and motion to constitute cheerfulness is manifested in several places laid out by Brown, where a lawn, surrounded by a sunk fence, and closed on two sides with corresponding rows of trees like blinkers, being left in a state of nature, but unoccupied by cattle, throws a veil of monotonous dulness over the scene, which no ray of cheerfulness can penetrate. Such was the case at Woolterton, in Norfolk, and at Kirtlington Park, near Woodstock.

If there be any truth in the above observations, it follows, that to plant out or remove such circumstances is a great mistake; and yet how frequently do we see a formal clump of larch or fir placed, either to hide a keeper's lodge, or to conceal a labourer's cottage, or to exclude the scattered hamlet, which we have been considering so essential, in some instances, to the cheerfulness of the scene.

Cassiobury affords a striking instance of

this mistaken planting. The Grand Junction Canal passes through the park, close under a high and finely wooded bank. Under such a circumstance, the improver, conceiving the canal an unsightly object, made a plantation of Larch, Fir, &c. to hide it : not perceiving that the consequence of hiding the canal would be the exclusion of the wooded bank beyond it— the finest feature of the scene. The plantation is now removed, and the occasional passing of the boats is a source of cheerfulness rather than of deformity.

If life and motion impart cheerfulness to scenery, cattle, and particularly sheep, should be admitted to the very boundary of the dress-ground. Nor should I be anxious to remove a pathway from the park, if it were not too close to the house : the occasional group of villagers supplying an additional embellishment to the landscape.

There are so many treatises on planting, as connected with soil, exposure, &c.—the result of greater experience than I possess on that subject,—that it is with due deference I venture an opinion as to the distance at which trees should be placed in forming a plantation. I cannot but think, that, under ordinary

circumstances, six feet is sufficiently near.
This will also assist the advantage of thinning,
which is seldom begun early enough.   I have
already suggested some hints on introducing
a considerable proportion of undergrowth in
all plantations, which will also materially
assist the thinning.   I will only observe, that,
if the spruce firs, as nurses, are cut off, when
about five feet high, they form, by the exten-
sion of the lateral branches, excellent cover
for game.   With regard to exposure, and
principally as it concerns the materials for
the dress-ground, I will mention a few circumstances which have fallen under my own
observation.   In the exposed situations in
Cornwall, the Ilex evidently stands foremost
in resisting the sea-air.   The Pinaster claims
the second place; but, though it resists the
blast for some time, it never, as far as I have
seen, becomes a tolerably good tree; whereas
the Ilex continues to flourish and improve in
size and foliage.   It should seem that the
silver fir stands the sea-breeze; as some of
the largest I ever saw are growing upon the
highest point of land at Tregothnan.   But,
not having met with them any where else,
under such circumstances, I can only state

the fact. I have, in one or two instances, found the cedar of Lebanon flourishing under nearly similar exposure. The sycamore is known to resist the sea better than any other deciduous tree. Among the shrubs, I have seen the Phillyrea most luxuriant under such exposure, and the Arbutus not far inferior. The common laurel shrinks beneath the saline atmosphere; the Portugal bears it better.

It has been already observed, that there is no tree which may not be advantageously employed in the decoration of scenery. I have ventured to condemn the Larch in park plantation; but, as a variety in the dress-ground, it is sometimes highly ornamental; especially if, from any cause, it has been diverted from its pyramidal form, or has lost its leader, and so assumed the character of Picturesque rather than that of Beauty. The most splendid example of the picturesque Larch I have seen is growing in the pleasure ground at Killymoon, in the county of Tyrone. The Larch may occasionally mix with good effect in a group. The grandest example of such a group is to be seen in the dress-ground at Wilton Park, near Beaconsfield, where two Larches of very large size and varied inclination entwine their elegantly sweeping branches with the more masculine arms of the spruce fir, and other trees of deep foliage, form a study worthy the pencil of Turner.

The mansion of early date is usually surrounded by trees of a corresponding age and size; which, as we have seen*, may frequently require partial removal, both for the improvement of the general composition of the scene, and that of the trees themselves, by throwing them into groups that shall produce a corresponding foreground. Trees thus connected with building are, I conceive, to be estimated rather by their appropriate character than by their intrinsic merit or individual beauty.

The cedar of Lebanon adorns alike the gayer lawn of the Grecian mansion, and the deeper recesses of the manorial pile; and rash indeed must be the hand that would remove it from either. But, had I the choice between the Oak and the Elm as accompaniments, especially to the manorial architecture, I should not hesitate to prefer the latter, as far more consonant with the general

* Page 48.

tone and sentiment of the building than the grandest Oak. Whoever has studied the perfect harmony that subsists between the " antique towers" of Eton College and the stately Elms which adorn its lawns will not hastily condemn this preference. The Oaks of Blythfield Park could not produce that solemn grandeur which results from the deep tone of colour and the monotonous masses of dense foliage of those Elms, whose grand, though simple outline, unbroken by any playful variety, are in unison with that contemplative solemnity, with which the scenc, and a consequent train of reflections, cannot fail to inspire the sensitive mind.

The Scotch fir has, of late years, been planted merely as nurse to the forest tree; but there are to be seen in many old places specimens which exhibit it in the very first rank of picturesque character, and approaching closely even to the grandeur of the cedar.

Various opinions have been suggested, as to the cause of this obvious declension of a tree, still continued to be planted throughout all the varieties of soil and climate in Great Britain. A very general mode of accounting for this declension is, that the closeness of the plantation not only prevents the expansion of the branches, but interferes with the life of the modern tree, as it is seldom found in a healthy state after sixty or seventy years' growth. This cause is not, however, of universal application, as very different results are observable under apparently similar circumstances. A considerable quantity of Scotch fir were lately cut out of the belt at Addington Park, not one of which had the smallest approximation to the old character; while there are now standing, in the belt at Drayton Manor, in Staffordshire, numerous specimens, which, though from a similar pressure they have lost their lateral branches, yet manifest, in the surface and colour of the bole, as well as in the rich luxuriance of the foliage, the true character of the Scotch fir. I have also observed the same circumstance in Kent, particularly in the neighbourhood of Tonbridge. The specimen here given was drawn from a tree of large dimensions, and in perfect health, though it must be of very considerable age; it stands amongst many others near the church at Sundridge.

A gentleman having mentioned to me some remarkable Scotch firs that had been

felled on an estate, I believe, in Lancashire,
did me the favour to write the following account of them :—

" I have made enquiries about the Scotch
" firs ; there were, so far as I can recollect,
" about twenty trees ; one tree was mea-
" sured in my presence ; it contained eighty
" feet of timber : on my observing what a
" magnificent tree it was, the carter said, he
" thought there was one a little larger. I
" should say, they were all nearly of the
" same size, certainly in height, but, perhaps,
" a little less timber in some.

" They had complete umbrella tops. They
" stood in a row ; and as there was one stood
" on the contrary side of the road, but lower,
" I should think there had formerly been an
" avenue of them. This last-mentioned tree
" was a poor stunted one ; perhaps not more
" than nine or ten feet of wood,—left because
" of little value.

" They were cut up for boards, &c. for
" farm and other buildings ; and the wood
" has proved to be of the soundest and best
" quality. It was so good, that a high price
" was offered for the wood by a timber mer-
" chant at Preston ; I forget the sum.

" I have stated all I can call to mind about
" them, and have spoken to my clerk of the
" works, who confirms my statement, as to
" the quality of the wood, and the magnifi-
" cent appearance of the trees, as he never
" saw any thing like them before."

Examples of this character, though of very
inferior magnitude, are to be seen in various
places; as in the long avenue on the road
between Dunchurch and Coventry, and near
the Litchfield race-course. The finest group
of them that I am acquainted with is in the
pleasure grounds at Teddesly, in Staffordshire.

It would appear that, from whatever cause,
the Scotch fir has, for many years, deteriorated
both in its picturesque character and in its
general estimation. An acute observer of the
beauties of Nature, more than half a century
ago, made the following remarks upon this
subject : —

" The Scotch fir is supposed to be the only
" indigenous Terebinthine tree in this island;
" and yet though it abounds, and when seen
" in perfection is a very picturesque tree, we
" have little idea of its beauty. It is generally
" treated with great contempt. It is a hardy

" plant, and therefore put to every servile office.
" If you wish to skreen your house from the
" south-west wind, plant Scotch firs, and
" plant them close and thick. If you want to
" shelter a nursery of young trees, plant
" Scotch firs; and, the phrase is, you may
" afterwards *weed them out*, as you please.
" This is ignominious. I wish not to rob
" society of these hardy services from the
" Scotch fir; nor do I mean to set it in com-
" petition with many of the trees of the forest,
" which, in their infant state, it is accustomed
" to shelter: all I mean is, to rescue it from
" the disgrace of being thought fit for nothing
" else, and to establish its character as a pic-
" turesque tree. For myself, I admire its
" foliage, both the colour of the leaf and its
" mode of growth. Its ramification, too, is
" irregular and beautiful, and not unlike that
" of the stone pine, which it resembles also
" in the easy sweep of its stem, and likewise
" in the colour of the bark, which is com-
" monly, as it attains age, of a rich reddish
" brown. The Scotch fir, indeed, in its strip-
" ling state, is less an object of beauty. Its
" pointed and spiry shoots, during the first
" years of its growth, are formal; and yet I

" have sometimes seen a good contrast pro-
" duced between its spiry points, and the
" round-headed oaks and elms in its neigh-
" bourhood. When I speak, however, of the
" Scotch fir as a beautiful individual, I con-
" ceive it, when it has outgrown all the more
" unpleasant circumstances of its youth —
" when it has completed its full age, — and
" when, like Ezekiel's cedar, it has formed its
" *head among the thick branches.* I may be
" singular in my attachment to the Scotch
" fir ; I know it has many enemies ; and that
" may perhaps induce me to be more com-
" passionate to it : however, I wish my opi-
" nion in its favour may weigh no more than
" the reasons I give to support it.

" The great contempt, indeed, in which
" the Scotch fir is commonly held, arises, I
" believe, from two causes.

" People object, first, to its colour. Its
" dark murky hue, we are told, is unpleasing.
" With regard to *colour in general*, I think I
" speak the language of painting when I
" assert that the picturesque eye makes little
" distinction in this matter. It has no attach-
" ment to one colour in preference to another;
" but considers the beauty of all colouring as

" resulting not from the colours themselves,
" but almost entirely from their harmony
" with other colours in their neighbourhood.
" So that as the fir tree is supported, com-
" bined, or stationed, it forms a pleasing tint,
" or a murky spot.

" A second source of that contempt, in
" which the Scotch fir is commonly held, is
" our rarely seeing it in a picturesque state.
" Scotch firs are seldom planted as *single*
" *trees*, or in a *judicious group;* but generally
" in *close compact bodies,* in thick array, which
" suffocates or cramps them ; and if they ever
" get loose from this bondage, they are
" already ruined ; — their lateral branches
" are gone, and their *stems* are drawn into
" poles, on which their heads appear stuck as
" on a centre. Whereas if the tree had
" grown in its natural state, all mischief had
" been prevented ; its stem would have taken
" an easy sweep; and its lateral branches,
" which naturally grow with as much beauti-
" ful irregularity as those of deciduous trees,
" would have hung loosely and negligently ;
" and the more so, as there is something
" peculiarly light and feathery in its foliage.
" I mean not to assert, that every Scotch fir,

" though in a natural state, would possess
" these beauties; but it would at least have
" the chance of other trees; and I have
" seen it, though indeed but rarely, in such a
" state as to equal in beauty the most elegant
" stone-pine.

" All trees, indeed, crowded together,
" naturally rise in perpendicular stems; but
" the fir has this peculiar disadvantage, that
" its lateral branches once injured, never shoot
" again. A grove of crowded saplings, elms,
" beeches, or almost of any deciduous trees,
" when thinned, will throw out some lateral
" branches, and in time recover a degree of
" beauty; but if the education of the fir has
" been neglected, he is lost for ever.

" Some of the most picturesque trees of
" this kind, perhaps, in England, are at Mr.
" Lenthall's deserted and ruinous mansion of
" Basilsleigh, in Berkshire. The soil is a
" deep but rich sand, which seems to be
" adapted to them. And as they are here at
" perfect liberty, they not only become large
" and noble trees, but expand themselves
" likewise in all the careless forms of nature.
" Very noble Scotch firs also may be seen at
" Thirkleby, near Thirsk, in Yorkshire. Nor

" has any man, I think, a right to depreciate
" the Scotch fir, till he has seen it in a per-
" fect state of nature."*

The author of these remarks, though he
regrets the contempt generally manifested
towards the Scotch fir, does not appear to
have been aware of the radical change which
I cannot but conceive to have taken place in
the plant itself. To account for this change,
as a means of restoring the original character,
would be a work worthy of those whose
knowledge and opportunities of investigation
fit them for the enquiry; and this not only
in a picturesque point of view, but also as
connected with profit, in providing an useful
and durable substitute for the worthless plant
now adopted.

Evelyn, in his Sylva, throws no light upon
this subject; but Dr. Hunter, in his edition
of that work, gives the copy of a letter from
Mr. James Farquharson, in which are the following remarks : —

" It is generally believed that there are two
" kinds of fir trees, the produce of Scotland,
" viz. the red or resinous large tree, of a fine

* Gilpin's Remarks on Forest Scenery.

" grain, and hard solid wood; the other, a
" white wooded fir, with a much smaller pro-
" portion of resin in it, of a coarser grain, and
" of a soft spongy nature ; it never comes to
" such a size, and is more liable to decay. At
" first appearance this would readily denote
" two distinct species; but I am convinced
" that all the trees in Scotland, under the
" denomination of Scotch fir, are the same;
" and that the difference of the quality of the
" wood, and size of the trees, are entirely
" owing to circumstances, such as climate,
" situation, and the soil they grow in.   The
" finest fir trees appear in the most mountain-
" ous parts of the Highlands of Scotland, in
" glens, or on sides of hills generally lying to
" a northerly aspect, and the soil of a hard
" gravelly consistence, being the natural pro-
" duce of these places.

" Upon cutting a tree over close at the
" root, I can venture to point out the exact
" age, which, in these old firs, comes to an
" amazing number of years. I lately pitched
" upon a tree of two feet and a half diameter,
" which is near the size of a planted fir of
" fifty years of age, and I counted exactly two
" hundred and fourteen circles or coats, which

" makes this natural fir above four times the
" age of the planted one."

It is to be regretted that the author of the
above observations, which were written in
1775, had not followed up this investigation
by such experiments as might, by this time,
have thrown a clearer light upon this interesting subject. The result of my own enquiries
amongst persons conversant with extensive
planting is, that the Scotch fir, properly so
called, is no longer propagated; but that the
tree now bearing that name was originally
imported from America; and the reason
assigned for its universal adoption is, that the
real tree gives out its seed with difficulty,
while that of the substitute is easily procured.

This idea appears to me to derive some
confirmation from a passage in the letter we
have just quoted, where the writer says, " In
" order to raise plantations of the Scotch fir,
" let the cones be gathered in the month of
" February, or March, from thriving young
" trees, as the old ones are not easily accessi-
" ble, nor so productive of seed. These are
" to be exposed to the heat of the sun, thinly
" spread on any kind of coarse canvass, taking
" them under cover in the night-time, and

" only exposing them when the sun shines.
" This soon makes the cones expand with a
" crackling noise. When any quantity of the
" seed is shed, it must be separated from the
" cones, otherwise the first dropped seeds
" would become too dry before the cones
" yielded their whole quantity, which often
" takes up a considerable time ; so that we
" are sometimes obliged to dry the cones in
" kilns, to make them give their contents in
" time for sowing — which ought to be done
" by the end of April or the beginning of
" May."

After all that has been advanced with
regard to the varieties of soil, climate, and
circumstances, being the causes of the declension of the Scotch fir, I confess myself unconvinced, while I find the remnants of the original character flourishing under all those
varieties in our own country ; and while the
broad distinction in the Baltic timber corro-
borates, in my opinion, a fundamental difference beyond what such varieties could
produce.

The Quarterly Review of Monteith's Planter's Guide laments, in common with myself,
the loss of the Scotch fir, though it does not

go into the details of the question. The
Stone Pine, in point of beauty, would supply
the loss of the old Scotch fir. Splendid specimens of this most picturesque tree are profusely scattered in the beautiful scenery of
Pains Hill, so justly celebrated in the Essays
on the Picturesque.

Where trees are to be planted as a foreground, and at the same time the view is to
be seen under them from the windows, such
trees should be selected as will bear pruning,
without destroying their beauty. Neither
the oak nor the elm are good subjects for
such a situation, which requires a more flowing line, and smoother bark than characterises
either of them. The beech, the ash, or the
sycamore, according to the circumstances of
soil and exposure, will be found equally applicable for this purpose. The choice among
them (where the above circumstances admit
of it) will be determined by the question,
whether a lighter, or a more massive foliage
suits best the general character of the scene?
And here, the improver, if he be unacquainted
with the study of landscape painting, will do
well to consult such pictures or prints as are
applicable to the subject; for, indifferent as

it may appear to the common observer, the effect of the composition is heightened or injured by a judicious or injudicious selection.

It would be difficult on this head to offer particular directions for general application, so much must depend upon variety of circumstances; but, perhaps, a few suggestions may assist in showing the advantage of applying the principles of landscape painting to the improvement of real scenery.

It may be sufficient for our present purpose to class the composition of landscape under the two leading characters of cheerfulness and grandeur. With the former of these, the elegant and pendant branches of the ash, or the light feathery extremities of the beech, are in unison: while the close formal outline and deep toned foliage of the sycamore assimilate with the latter. It will be obvious to the most casual observer, that many circumstances of difference in each of these characters of landscape, and many more arising from the varied degrees of mixture with each other, will demand various modifications of the above hint. The solemnity of the sycamore may be relieved by the light playful birch, or the more masculine, yet elegant limbs of the Spanish chestnut, which ought, indeed, to have been named as a principal, rather than an auxiliary, in forming such a group, as no tree offers a fairer subject for the pruning necessary for it. The cheerful character may, in like manner, occasionally require a mixture of closer foliage and deeper colour. Variety is the leading feature of the cheerful: unity the characteristic of the grand. An example of a foreground group of the lighter character will be found at Woollaton near Nottingham, composed of noble shafts of beech: a splendid specimen of the grand may be seen at Killymoon, before mentioned, where a group of sycamore of gigantic size, standing on a jutting knoll, makes the foreground of a most romantic composition, formed by the junction of two brawling streams, struggling together in a rude rocky channel, overhung by high woody banks. The catches of light playing on the agitated water, contrasted with the deep sombre tone of the surrounding scenery, reminds you of the happiest effort of Ruysdale's pencil.

In planting groups of trees, the number should not be the same in each group; a thorn or two, occasionally introduced, gives variety to the character. When the group is composed of three trees, two of them, in my opinion, ought to be of one kind, as three distinct ones can hardly be supposed the result of natural combination, at which all planting should aim. For this reason, I should always plant two or more trees of one kind on those points of plantation, which are hereafter to be separated as a group. With the same view of imitating the accident of Nature, trees should not be set at equal distances. Two may be nearer, and one considerably more remote. Frequently two may stand close together; but, in that case, a fourth will generally be necessary to give a right balance to the group. As the trees composing the groups should not be at equal distances, so neither should the groups themselves. The character of the ground, the situation of some mass of wood, or some other local circumstance, will, if attended to, frequently suggest occasions for variety in their disposal.

It may here be observed, that a woodman should never be admitted among trees designed for ornament. *His* aim is to create individual distinctness; that of *the landscape painter* to promote the intricacy and variety of composition.

The practice of levelling the surface has done much mischief both in park and pleasure ground. No one, conversant with the study of landscape, can fail to observe and to regret its baneful influence in Hyde Park, where smoothness and clearing have been deemed essential accompaniments of the drive. Nor is this, in my opinion, the only mistake in the late embellishments of this park. The striking scene, which meets the eye on entering from Piccadilly, will be blotted out by the line of trees planted along the ride. They would have been better placed on the other side of it. As they are now situated, the ample lawn, enriched with varied groups of fine trees, and backed by the bold line of Kensington Garden wood, the whole smiling cheerfully on the beholder, will shortly be changed into the monotony of a row of trees, excluding all variety of form or colour; and this will be the dull attendant of the ride.

But to return to the subject of levelling. Hardly is there a more common mistake, than that of filling up a gravel pit, should it be within sight of the approach or the drive;

whereas, there are few circumstances more
capable of decoration; especially if accompanied with water, which is frequently the
case where the pit is of sufficient depth. The
bays and promontories, with all the acci-
dental varieties of excavation, afford excellent
models for the formation of artificial water;
and, with judicious planting, the pit itself
becomes a picturesque appendage to the
scene. A fine example of this kind of decoration may be seen at Dunmore Park, in
Scotland, where the stone quarry, from which
the mansion was built, exhibits a composition
truly romantic. A rocky precipice, crowned
with trees, and reflected in the pool below,
whose sides are fringed with wood, is caught
under various combinations from a path, that
winds its devious way, through the inequa-
lities of the excavation, from the bottom to
the top of the quarry.

When, from any circumstance, spare earth
is to be disposed of in the pleasure ground,
it is usually applied to the filling up of any
hollows that may fortunately exist: whereas
it should be used to increase any indications
of undulating forms; as even the smallest
variety of this kind is highly advantageous,

whether in the lawn, or in the plantations of shrubs which surround it. It will be safer for the unpractised eye to increase the existing varieties of the ground, rather than to create new ones; the arrangement of earth for this latter purpose being an operation of considerable difficulty; whereas a moderate degree of caution cannot well fail in the former.

The fencing of plantations is a subject of considerable difficulty, arising principally from the expense attending it. The general method of a quick fence upon a bank, when applied to plantations near the house or the approach, is highly objectionable, as it excludes the boles of the trees, whether the hedge be suffered to grow wild or be kept clipped.

The most desirable fence is, doubtless, that which is least observable. For this purpose, iron hurdles have sometimes been used; but, where the plantations are extensive, the expense becomes a serious objection.

I have lately seen, in several places, a wire fence, which appears to me likely to reconcile the contending objects of beauty and expense; for I am informed, that it can be put up at the cost of from fifteen to eighteen pence a yard, and that it will resist heavy

P

cattle. In situations where sheep only are admitted, no doubt of its sufficiency can be entertained.

If such a fence can be obtained at the cost mentioned above, I should conceive it to be cheaper than a quick hedge, as that must itself be protected by a post and rail till it becomes capable of resisting the stock.

A friend of mine has put up a fence described in the sketch at the cost of a shilling a yard, upon the following plan : — Wooden or iron standards placed at forty yards apart, with iron uprights a rood distance from each other. The uprights are not fastened into stone or wooden blocks, but merely driven eighteen inches into the ground. A notch, such as is shown in the example, is made at the top and bottom of the upright, the former about two inches from the upper end, the latter six inches from the ground, so as to prevent lambs from getting under the wire. The wires are confined in these notches by thinner wires twisted round them, which are also passed round the intervening lines of wire, as represented in the example. I am not aware of any advantage in this mode above the usual method of drilling holes in the uprights, except that it greatly facilitates the changing of the direction of the fence, which, in the situation to which we are applying it, will not be required. If, after all, the wire fence should be deemed too expensive for large plantations, it may be used for those parts which are near to the house, or to the approach; while the less visible portions may be protected by a hedge, with as little bank as possible. It is not necessary that the fence should follow all the varieties in the outline of the plantation; the lesser angles of lawn may be filled with furze, fern, or brambles, which all assimilate with the wood.

Where the occasional thinnings are sufficient to supply a fence as long as protection is necessary, I should prefer a post and rail to a hedge, as the boles of the trees, and the recesses in the plantation, will be visible through it.

It may not be amiss to state, that a deer fence surrounding a plantation is not required to be of a greater height than one for general purposes, as, except they are driven, the deer will not leap into such an inclosure. The fence of the new plantations in Richmond

Park is only four feet high; and that at Cassiobury the same.

Where a quick hedge is made the fence to a plantation, it should follow the varieties of the outline; otherwise, those varieties will be lost. The effect will be considerably assisted by groups of thorns planted at different distances from the hedge itself; thereby relieving its uniformity of surface.

The adaptation of the entrance lodge to the residence has occasioned much discussion, without leading to any fixed principle. It is not with any pretension to adjust this difficulty that I offer a few remarks upon the subject; but, principally, to suggest a consideration, which I do not recollect either to have met with in print, or to have heard *vivâ voce.*

Among the various opinions on the propriety of a lodge, and the numerous examples for its different styles, the subject has been considered merely with relation to the residence to which it is attached.

Now, I conceive the question of propriety to depend at least as much upon the character of the scenery where the lodge is placed as upon that of the house which it accompanies.

The splendid gateways at Burleigh and at Woburn, opening into the grand and extensive scenery of the parks, are perfectly in unison with that scenery ; but were any approach to enter the domains at some spot where the inequality of the ground, and the confined scenery, afforded little room for display, I should conceive such gateways sadly misplaced. A gamekeeper's cottage would be more in harmony with the scene, and therefore, in my opinion, every way more appropriate.

Dalmeney, in the neighbourhood of Edinburgh, affords a good illustration of this idea. The mansion is worthy of the extensive and beautiful domain over which it presides ; but the shape and character of the ground at both the entrances forbids any attempt to the erection of a lodge corresponding with the architecture of the house ; and, I think, good taste has substituted a simple building at each of them.

Where a splendid lodge is appropriate, I should prefer an arched gateway to any other building, provided the gateway be deep enough. A thin archway, being deficient in point of light and shadow, is to me, I confess, very unsatisfactory. Moreover, this depth admits of the porter's residence within it, which I take to be more simple and characteristic than any separate lodge united with such a gateway can be. A wall of corresponding height and character should accompany this entrance.

A gate between pillars, if upon a large scale, seems to require a lodge for each flank; but such an arrangement, appearing as a sacrifice of comfort for display, is not, perhaps, in the best taste. A pair of lodges under any circumstances of less grandeur, are, in my opinion, utterly indefensible.

The numerous designs, given for what are termed *Gothic lodges*, afford ample room for choice; which choice should depend upon the circumstances of each place. I will only advise, that simplicity of composition direct the choice; as I cannot but feel, that many of these designs are over-done in chimneys, gables, porches, &c.

I should recommend the improving of any cottage that may happen to stand near an entrance gate (if capable of improvement), in preference to the erection of a Gothic lodge. The addition of a porch to the door, of hoods to the windows, and other little decorations, will, under good taste, often produce a picturesque lodge, pleasing on account both of its simplicity and its rarity, —as the Gothic is found at almost every entrance.

An example of such a lodge may be seen at Bickley, near Bromley, decorated by the same hand that erected those beautiful specimens of the Gothic cottage at Redleaf.

The gate will vary with the character of the lodge. If the arch gateway be of the Grecian architecture, an iron gate seems the most appropriate: but then it should be massive and rich in its construction; and, in my opinion, should fill the whole arch. If the gateway be of the Gothic character, I should prefer a close wooden gate as better suited to the sombre tone of the building. This gate should range with the spring of the arch, and be a straight line on the top. A gate with open bars half way down is not unsuitable to the Gothic, provided the bars are massive, and the mouldings bold. The colour of the wooden gate should be that of oak. I should advise a straight line for a gate in all cases, except where it fills the arch.

As I conceive a lodge to be governed more by its relative situation than by the mansion, so also, in my opinion, the situation is the primary object to be considered in the character of a bridge.

Where the scenery warrants, and the splendour of the mansion demands, an architectural bridge, if I may so speak, the degree of magnificence or decoration will depend upon the degree of those qualities exhibited in the mansion itself. In all cases it should be horizontal.

It is not, however, often that the width of the water, or its proximity to the house, demands such a bridge. Where the scenery justifies such a structure, it forms a beautiful feature in the landscape, as at Clumber.

In general, a bridge is required merely to cross some rivulet or brook, which interrupts the approach; and under such circumstances, whatever be the extent and magnificence of the domain and the mansion, picturesque effect should prescribe the character of the bridge.

Utility being the primary object in all

simple structures, ornament is at best misplaced when applied to unadorned nature. Upon this principle, I prefer a plain wooden bridge for the crossing of a shallow stream, to which a regular arch seems superfluous. The stays of the annexed sketch, while they give a variety to the outline of the bridge, yet are all apparently necessary to its stability.

Should the stream require a longer bridge, arches will become necessary; and they will be built of brick or stone, as circumstances may dictate. But, though masonry will form the basis, I should prefer having the battlements of timber, which should overhang the masonry beneath. Many examples of such construction are to met with in unfrequented country roads.

The same principle will apply to footbridges, as to those we have been considering. The beautiful Palladian bridge at Wilton is in perfect harmony with the mansion, and with the magnificent cedars which accompany it; but were a wood-walk in the same domain interrupted by a stream, the bridge should be of the simplest character, in compliance with the scene.

I am not fond of what is termed *a rustic
bridge*, as lightness I conceive to be essential
to such a structure. Neither would I have
an iron bridge in such a situation, as it wears
too dressed an appearance.

It is desirable, that in proportion to the
size of the domain, its influence in improvement should be extended to the scenery
around it; which scenery will frequently
include the village.

Villages may be divided into the regular
and irregular. The irregular village is doubtless the more picturesque, admitting of greater
variety of composition from the intricacy of
its outline.

The improvement of such a village (as far
as picturesque effect is concerned) need not
be expensive: the principal part, in many
instances, will consist in preserving the varieties of form and material existing in the
different cottages that compose it. A porch
where it may be required, hoods occasionally
placed over the windows, will relieve the flat
surface of the wall, and add those deep
shadows so necessary to effect. A wall, a
hedge, or a paling to be restored, will com-

prise all that the greater part of such a village will require.

Care should be taken to preserve all those varieties of outline, those irregular and angular projections, which mark the old houses; and if it be necessary to rebuild the chimneys of such a structure, the original character should be adhered to.

Neatness being an essential characteristic of a village bordering upon a mansion, will require the hedge that bounds the cottage-garden to be kept nicely clipped; and the garden itself to manifest in its keeping the pleasure it affords to the inhabitant. Creepers of various kinds will adorn the porch, and dress the wall or paling, if such be the fence; while flowers and shrubs will decorate such part of the ground as can be spared from culinary purposes. The holyhock, in its varied and luxuriant hues, rising high above the simple fence, and breaking the quiet tone of the building, is peculiarly adapted to such a garden. In fine, the decorations of the various dwellings should appear rather as the result of the feeling of each inhabitant, than as arising from any regular plan of improvement.

One great requisite in village scenery consists in trees : indeed, the village is not perfect without such accompaniment. What
beautiful examples do we occasionally meet
with !—A yew, sheltering with its dense foliage the projecting porch, and relieving by its
sombre tone the light and playful creeper
which adorns the rustic tracery, — a cypress,
or a Virginia cedar, contrasting the horizontal
lines of the roof and eves of the cottage, its
head mingling in group with the shaft of the
chimneys above. Every opportunity, therefore, should be taken to enrich the general
composition by this accompaniment of foliage;
adapting the tree to its situation, both as to
size, colour, and character.

As variety in colour, as well as in form, is
essential to picturesque effect, great caution
should be observed on that head of improvement. Indeed, village architecture, almost
over the whole of England, is suffering as
much from the colour as it is from the form
of modern cottages. Wherever it is necessary to rebuild a village, brick is nearly the
universal material ; and if its fiery tone should
be deemed offensive, a general whitewash is
the remedy. How jealous, then, should we be

to preserve such varieties as the old village
may afford. The rich varied tints of an
ancient house ought never to be interfered
with; and should the walls of another be
unsightly smeared, it should be restored by a
sober wash, approaching to drab rather than
to white; another might be of a grayer tendency; but pure white scarcely ever can have
a good effect. Sir Uvedale Price objects to
thatch, from its damp and dirty look when
passing into decay; but I consider it, almost
under any circumstances, to be preferable to
tiles, or the cold blue slate generally used.
The flat tile, when old, becomes picturesque,
and well adapted to the village roof.

Where a mansion is so unfortunate as to
have a regular village for its neighbour, all
attempt at the picturesque is hopeless, especially if there are houses on both sides of the
road. Neatness, and a degree of symmetry,
are all that can be effected. Ripley, near Har-
rowgate, is a good example of such a village.

When a village is to be partially rebuilt,
every opportunity of irregularity should be
attended to; and care taken to assimilate the
new, as far as convenience will permit, to the
old. Any swelling knoll that will contain a

cottage should be so occupied, as tending to that variety we have been considering.

A village sometimes appears as the residence of people above the labouring class, from the size and decoration of the houses which compose it. The village of Belton, near Grantham, exhibits a striking example of this character: the variety of the houses, and the picturesque form of each, evince the taste of the designer. The highest specimen of the straggling village with which I am acquainted is adjoining the grounds at Foxley, the residence of the late Sir Uvedale Price.

Being very little conversant with flowers, as to their varieties, culture, &c. I merely mention a circumstance with regard to the dahlia, which strikes me as a great improvement in that universal favourite. The flower of the dahlia is doubtless a magnificent display of beauty and variety; but, from the quantity of coarse green leaves, and the height of the plant, it never appears to me to form agreeably in a bed, either alone or mixed with other flowers. A person mentioned to me his having seen it produce a splendid appearance from each plant being pegged down when about a foot high; and thus covering the whole bed with flowers. I merely repeat what I was told, not having had any opportunity of seeing the effect.

As a solid walk is in all cases an essential comfort, and gravel in many places difficult to procure, the following method of supplying that want may be serviceable as a substitute for that useful article. As in the instance of the dahlia, I only repeat what was told me, never having seen the experiment tried.

To make paths quite hard:—Rough gravel, or broken stone, at bottom, about three inches thick; to be covered from two and a half to three inches thick with a composition of fine beach or coarse sand and fine chalk rubbish, in the proportion of one of fine chalk to three of sand, made up with water and gas tar (using one of tar to six of water) and beat down solid to a fair surface.

In the search after truth in any science, it has ever been held allowable to examine freely such publications as have treated on the subject in discussion.

Upon the same principle, I trust I may be permitted to elucidate my ideas upon the subject of taste by a reference to such public works as are connected with that subject; as it is only by examples of notoriety that the question can be fairly brought to an issue, so as to assist in the establishment of some standard on a question of such general and increasing interest.

It is under this sanction, that I have ventured to give my opinion upon the treatment of the water at Buckingham Palace, and also upon the late improvements in Hyde Park; and, under the same sanction, I now proceed to offer a few remarks upon another public work, which is closely linked with the subject of these pages.

In treating of the terrace, I have held a strait line to be an indispensable characteristic. Under this impression, I cannot but view the terrace round the flower garden, at Windsor Castle, as a mistake. The curving line, in my opinion, destroys the very essence of a terrace. It should have been a strait line, parallel with the Castle front to which it is attached; and it should have been entered upon by a flight of steps at each end.

Another advantage would have been obtained by forming the terrace on a strait line; the whole of the wall surrounding the flower garden would have been angular, which I think would be more in harmony with the Castle than the present form; and it would, moreover, have given an opportunity for a pavilion at each end of the terrace, and thus relieve the continued flatness along so great an extent of masonry.

This flatness I should now endeavour to break by shrubs planted on the banks above the flower beds, which, in my opinion, would be a great improvement to the general composition of the scene, as tending to promote that richness and variety so essential to the size of the garden. Such a process would also apparently lessen the depth of the flower beds below the terrace, which I cannot but think rather too great.

Having already expressed my opinion, that it is essential that a professor should explain the principles upon which he suggests any improvement, I would now warn the proprietor not hastily to adopt any plan which cannot be thus explained; as, I am sorry to say, I have seen too many instances of irreparable mischief arising from the utter ignorance of the professional improver.

Alteration is frequently mistaken for improvement, and laborious operation for superior taste. I should recommend caution in adopting any proposal of the latter class, as a professor destitute of true principles will overlook those apparently little circumstances upon which real improvement frequently rests, and will proceed to a total subversion of the scene which he knows not how to adorn.

If the practical hints I have thrown together have any foundation on principle, I trust they will assist the proprietor in determining the character of his own place, and in adapting his improvements to that character. This is the first lesson to be learned, without which, if any real, though partial improvement is effected, it must arise from accident.

I will offer another observation, which concerns equally the proprietor and the professional man.

It will occasionally happen, that the proprietor does not embrace the whole of the proposed plan, but does not state his objections at the time. The consequence will sometimes be the mutilation of that which, had the objection been stated in the first instance, might have been so managed as to meet the proprietor's ideas, without deranging the whole scheme.

In justice to the professional improver, I will beg leave to observe, that he is not to be held answerable for the discrepancies that will occasionally be found between his principles and the illustration of them in various places.

Besides the mutilation above mentioned, instances will occur of the most flagrant violation of every principle of taste.

I was consulted, some years ago, upon the improvement of a place worthy of every attention. The house was undergoing a repair, and a drawing-room, or library, was to be added, commanding a fine view over a varied landscape.

The original approach came too near the end of the house where the additional room was to be built. The hall door was where it ought to be — remote from the living rooms. An old avenue presented an opportunity for a fine approach.

These circumstances I pointed out to the proprietor, and rejoiced in the prospect of getting rid of the road from the living rooms, and in so fortunate a line as the avenue offered for it.

I left the country, anticipating the beauty

of the dress-ground, and the scenery beyond it, when relieved from the intrusion of the approach. What was, then, my astonishment and mortification when I learned, a short time ago, that the hall-door was placed close to the library, and the approach carried under the windows.

This fact, among many others, will evince, that local advantages, aided by the happiest plan of improvement, are of no avail when a false idea has taken possession of the mind. The reason given for the irreparable error which I have stated, is in itself contrary to the first principles of taste, viz. that the approach should exhibit the beautiful scenery, which ought to have been reserved for the windows and the dress-ground.

Should these few pages be the means of correcting only this one error, they will not have been written in vain.

THE END.

LONDON:
Printed by A. & R. Spottiswoode,
New-Street-Square.

.

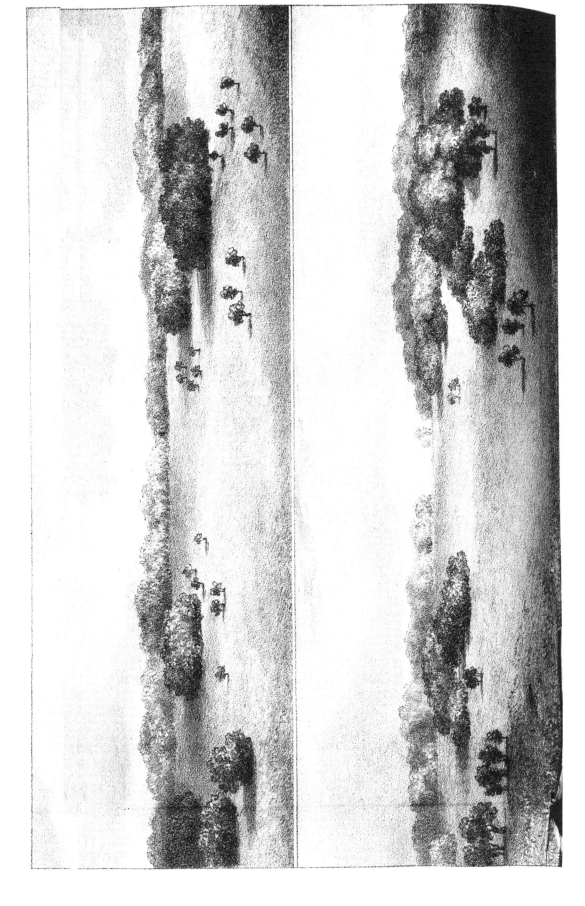

These uprights a rod apart. The wires N°4 and N°5, the thickest for the two upper lines. The lowest wire 6 inches from the ground.

The upright 4ᵗ high from the ground

18 inches in the ground

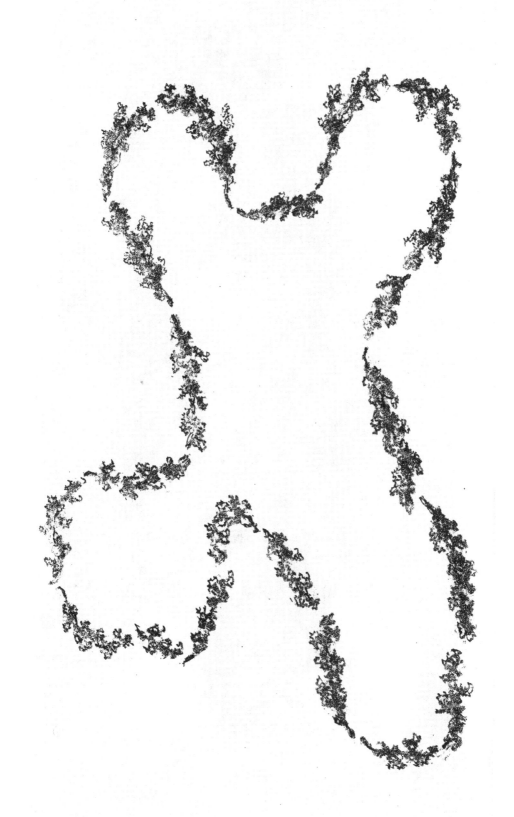

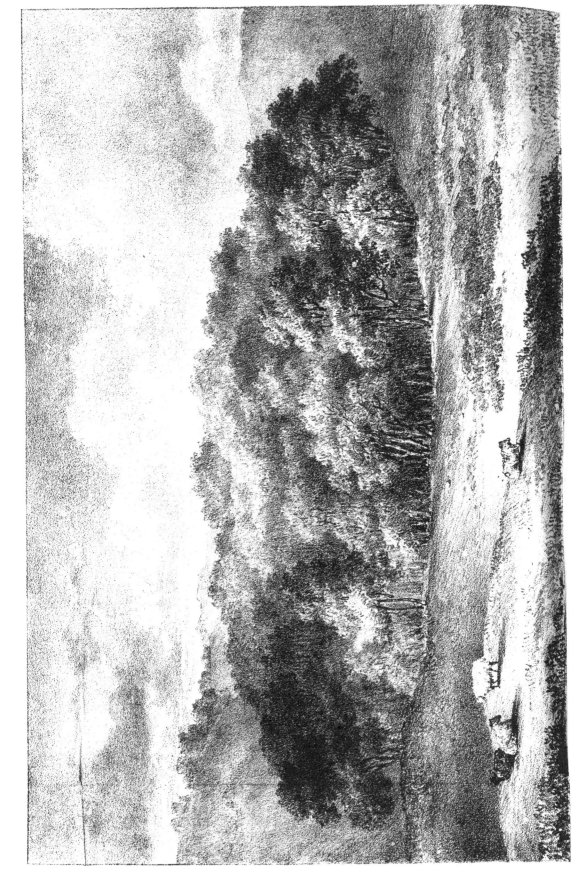

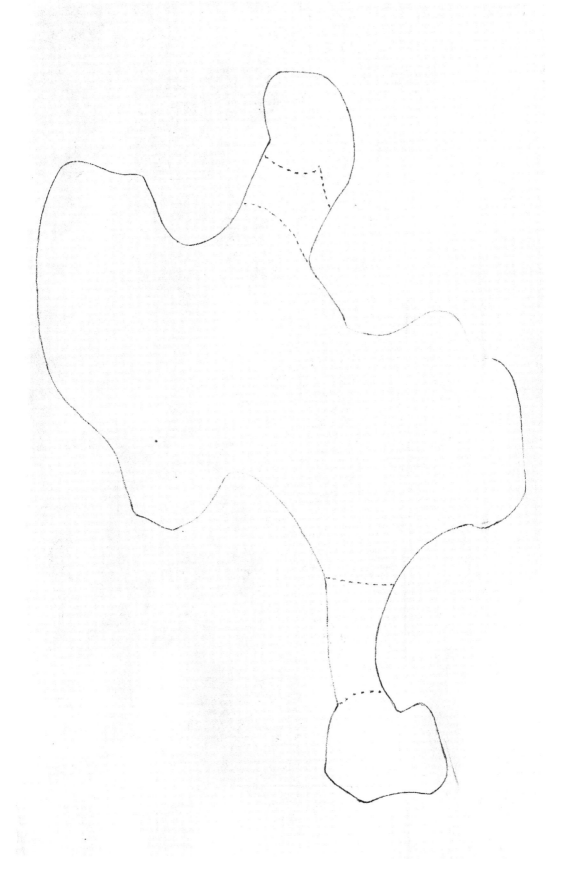